Checked-In

How To Use Gowalla, Foursquare and Other Geo-Location Applications For Fun and Profit

By Herbert Tabin and Craig Agranoff

Published By Pendant Publishing – www.pendantpublishing.net

Copy Editor: Aaron Turpen

Book Designer and Compositor: Rosamond Grupp

ISBN: 978-0-578-06348-5

Printing History – First Edition: August, 2010

Foreword

For the last decade or so, the idea of "life as a virtual game" had been discussed about in tech circles without results – until recently.

All over the web, reward-based achievement games, through geotagging (the process of linking content like text, images or video to a specific location identified by GPS) have begun to arise as a means of encouraging specific behavior. Now, after years of hype, people are using their phones to tell others willingly exactly where they are, and marketers are lined up for a piece of this "location targeted consumer" action. So it is now that geolocation is starting to be used to connect people in the physical world. It's a marketers' dream: real-time social networking that can tag people's eating and shopping preferences.

WIth geolocation, no longer will advertisers see their marketing dollars wasted for example by showing ads for restaurants to people that just finished lunch, instead, advertisers can now know exactly who their customers are – what they are doing, where there doing it, at any given time.

So who are the leaders hailed as the next big thing? Foursquare and Gowalla. Both are gaining traction quickly. They are setting the stage for location-aware future full of "local promotions" targeting their active user base it's about the services that are pioneering and revolutionizing the way we will shop, find friends, play games and yes, sell products.

So if you want to understand that huge revolution taking place right now and take part of it from a user level to better understand the concept, this book is for you.

Martin Thorborg
Co-founder SPAMfighter.com and author of ePusher

About The Authors

Herbert Tabin is an author consultant, entrepreneur, venture capitalist, merger and acquisition expert as well as a noted specialist in online reputation management and monitoring. He is the author of "**Do It Yourself Online Reputation Management-** *A Step-By-Step Guide to Building or Repairing Your Online Reputation*" and has worked with many Internet startups founding Vois.com, the blog sCommerce.com and operated Rev2.org. Tabin has been featured on the cover of Entrepreneur Magazine - Entrepreneurs Business Start-Ups, The Palm Beach Post, and in Steve Spalding's book – 'All the Little Things' – Get the advice you will need to get your idea out of the garage and onto the Web. He is a frequent speaker at technology events such as Refresh and LaidOffCamp.

Over the years he has received various entrepreneurial award nominations including Ernst and Young's Entrepreneur of the Year Award, receiving The Award for Business Leadership. He was also nominated as a finalist for best Large Scale Social Network, winning best "Photo Sharing Website Category" at Mashable's™ Open Web Awards. In March 2000, the State University of New York At Oneonta named its largest computer lab, the 'Tabin Computer Lab'.

Craig Agranoff is an entrepreneur, and national Social Media consultant as well as a noted specialist in online reputation management and monitoring. He is the author of "**Do It Yourself Online Reputation Management**- *A Step-By-Step Guide to Building or Repairing Your Online Reputation*" and has worked with many Internet startups founding the tech blog sCommerce.com and was the editor or Rev2.org. He also is a Tech/Social Media Correspondent in the New Times, VentureBeat.com, and Reviews.Vois.

Craig Agranoff has been featured in the Miami Herald, New Times, The Palm Beach Post, AOL Digital Cities, Slice, CenterNetworks, Thrillist and The Sun-Sentinel. Agranoff has also appeared on Gary Vaynerchuck's Wine Library TV, Fox News, and CBS News.

He is a frequent speaker at tech events such as Refresh, New Tech Community, Bar Camp and LaidOffCamp. He was also nominated at Mashable's Open Web Awards as a finalist for best Large Scale Social Network, winning best "Photo Sharing Website Category".

Additionally, in his spare time, Craig's the Pizza Expert and founder of WorstPizza.com.

Table Of Contents

Introduction

It all began with Twitter in 2006. With Twitters' micro-blogging service people quickly learned to communicate their every thought with others in just seconds. Whenever an author wanted to send a message to their friends, associates and other subscribers (known as *followers)* Twitter "tweets", or text-based posts of up to 140 characters, were delivered. From mere obscurity just a few years before Twitter now ranks as one of the 50 most popular websites worldwide by Alexa's Web traffic analysis with its' members currently posting more than four billion tweets a quarter. What began as simply the "what I am doing now" phenomenon has begun a new phase spilling over into "where I am now".

"Where I am now" is based on a person's exact geographic location using Global Positioning Software or GPS. With smart phones this "geo location" trend has spurred a spade of new websites and applications with the leaders being Foursquare.com and Gowalla.com. Other players include Britekite.com, Whrrl, MyTown, and the UK's Rummble.com.

With this book, you will learn not only how these geo-locations applications grew into their new-found popularity, but also how you can use them yourself for fun, finding new friends and even promoting your business. Geo-location is poised to become the next big revolution online and if you want to be a part of it, you'll need to know how to utilize these apps.

Let's see how you can tap the power of geo-location.

Chapter 1

Geo-Location, Gowalla and Foursquare

What is Geo-Location?

Geo-location is the identification of the geographic location of a physical computer or electronic device that is connected to the Internet. What began as the general area of the device's location based on where it connected to the Internet eventually grew to incorporate Global Positioning System (GPS) coordinates as hand held devices became popular.

With the introduction of GPS to the Apple iPhone and Android operating system of smart phones, geo-location is becoming wide spread. Prior to this introduction, geographic advertising was already gaining ground as an evolution in online sales, based on simpler, "general area" locations rather than *specific* locations on the map (e.g. "Denver, Colorado" rather than "4th Street and Broadway, Denver, CO").

Although smaller geo-location-based games had existed before, most notably "Geo Caching," (http://www.geocaching.com) until smart phones and GPS combined, the stage had not been set for this new revolution to take place.
Soon after GPS became common on smart phone platforms, startups began using it as a way to create new marketing, content, and social networking services on the Web. Website pioneer Gowalla.com expanded to a national platform followed quickly by the introduction of Foursquare.com.

What is Gowalla.com?

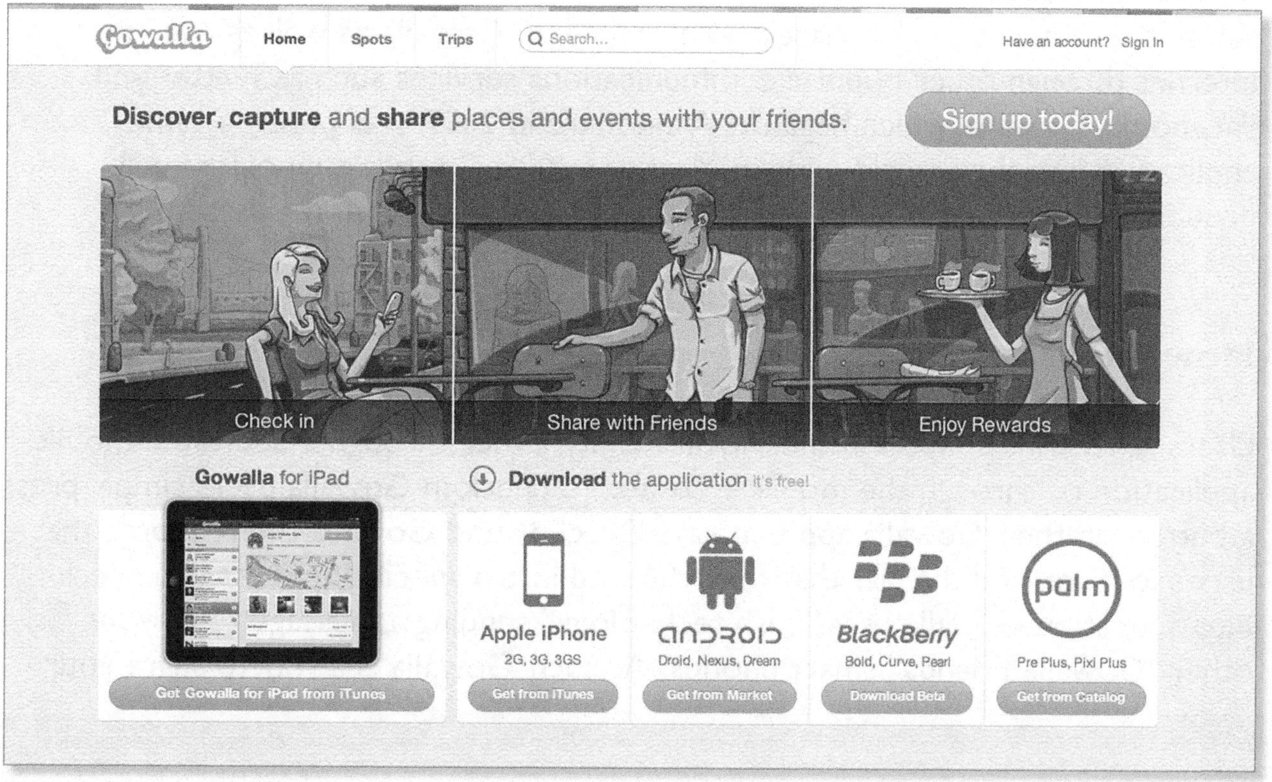

Gowalla.com or "Gowalla" is a location-based travel game that uses either an iPhone or Android device to reward both everyday and unusual location visits. These experiences can be shared with friends on Gowalla as well as over the internet through other popular communications services such as Twitter and Facebook. Gowalla, which began as Alamonfire in 2007, was purely a game application launched world-wide in March of 2009. Gowalla supporters will proudly tell you, Gowalla was geo-location before geo-location existed.

How Gowalla Works

Users "check-in" at *Spots* in their local vicinity, either through a dedicated mobile application or through the mobile website. To check-in Gowalla users simply press a button on their Gowalla app that says Check In and Gowalla then records that user's location and the fact that they checked into a specific Spot. As you check into a location Gowalla uses "push technology" sending a message to all your other "Gowalla Friends" smart phones. As such, Gowalla is a way to share your current location with your friends so you can sometimes have random meet ups or get recommendations on places to visit while you're there. Marking yourself as being at "18th Street and Main" might tell you that a friend is in the coffee shop two doors down or, after alerting your friends to your location, have them recommend that coffee shop to you.

By using the Gowalla mobile app, you can stamp your virtual passport (within the Gowalla application) when you arrive at a location, such as your favorite coffee shop or even your own office. If the location hasn't been "discovered" by another Gowalla user already, you can create it, no matter where in the world it is. Some locations will contain bonus virtual items "stashed" (or "geo cached") in them that you are rewarded with when you arrive as a reward for 'checking in. These items are either randomly placed or can be placed there out of the inventory of a prior visitor to be found by the next visitor. These items are images of things like a "yellow bicycle" or a "pint of beer" and are virtual items you carry in your Gowalla inventory.

AT&T 3G
3:31 PM
Items
Pair of Nets Tick...
Pair of Nets Tickets
ISSUE NUMBER
86 of 500
ADMIT 1
Add to Collection
This item is missing from your collection.

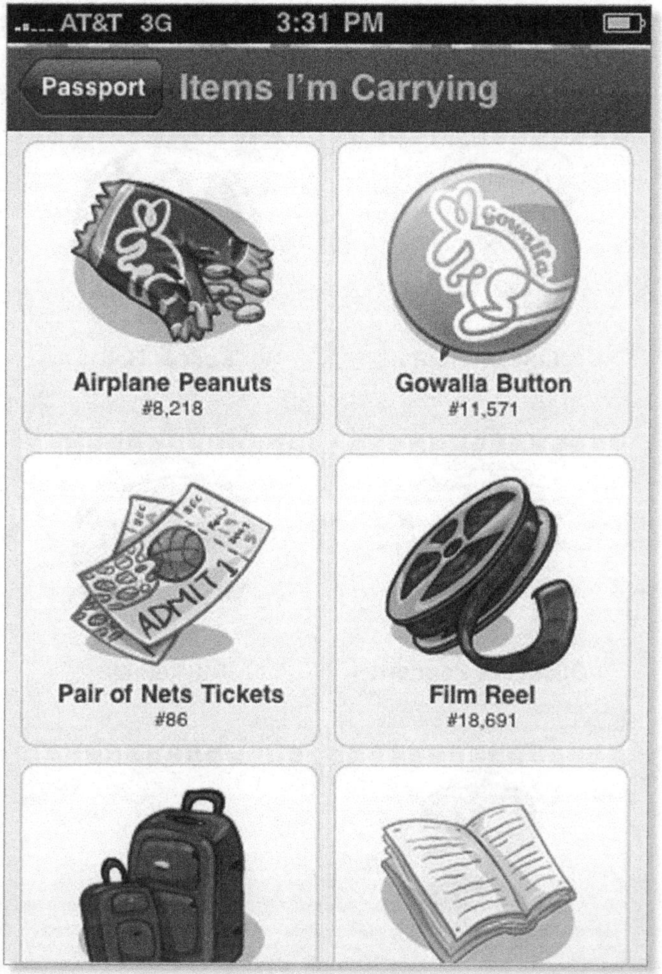
AT&T 3G
3:31 PM
Passport
Items I'm Carrying
Airplane Peanuts
#8,218
Gowalla Button
#11,571
Pair of Nets Tickets
#86
Film Reel
#18,691

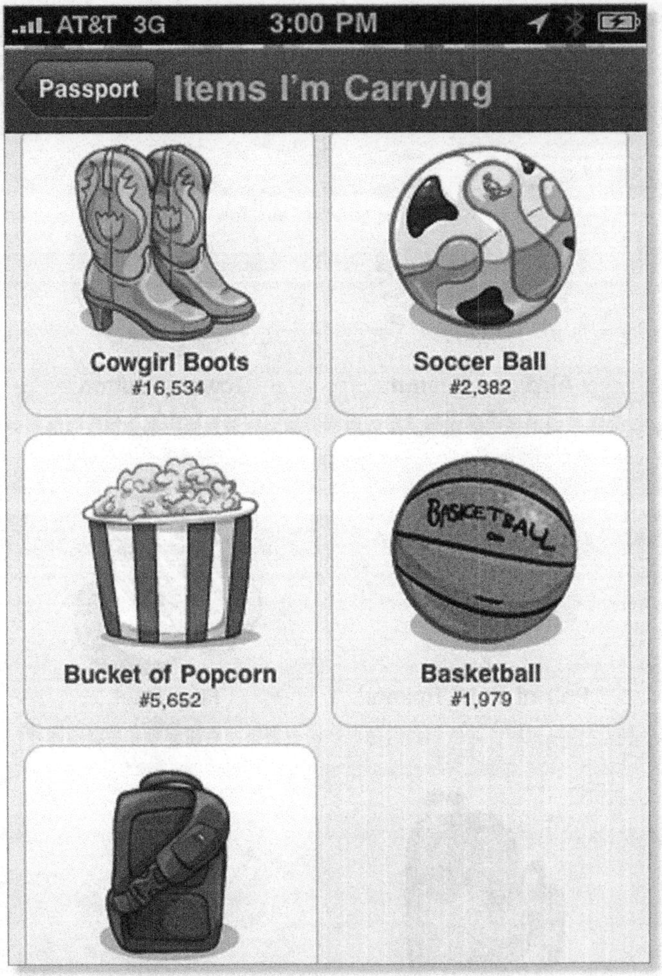
AT&T 3G
3:00 PM
Passport
Items I'm Carrying
Cowgirl Boots
#16,534
Soccer Ball
#2,382
Bucket of Popcorn
#5,652
Basketball
#1,979

As you collect these virtual items, you can show them off to your friends showing how many you have acquired and if you want you may share them with others by leaving them behind for others to locate as well. Each item has a history that you can view to see who's held the item before and what locations it's been in. Some, items have literally traveled the world!

Entrepreneurs and creative business owners are beginning to see some real possibilities with Gowalla and users are sometimes finding that they get real-world rewards for using the service at some locations. We will get more into this aspect later in the book.

What is Foursquare?

Foursquare is a geo-location application that combines geo-location sharing with social networking and even online gaming. It connects you with your Foursquare friends as well as social networks like Twitter and Facebook.

Foursquare was founded in 2009, and started as a simple geo-location-based game concept that grew quickly in popularity. Foursquare now rivals the older Gowalla in users for supremacy in this fast-growing niche.

At it's core, Foursquare is a way to 'check-in' and share your current location with your friends so you can sometimes have random meet ups or get recommendations on places to visit while you're there. Marking yourself as being at "5th Street and Junction" might tell you that a friend is in the cafe two doors down or, after alerting your friends to your location, have them recommend that cafe to you.

On top of this, Foursquare has a little politicking (in a game-like fashion, of course). The Foresquare difference is that every time you check in at a location, be it in

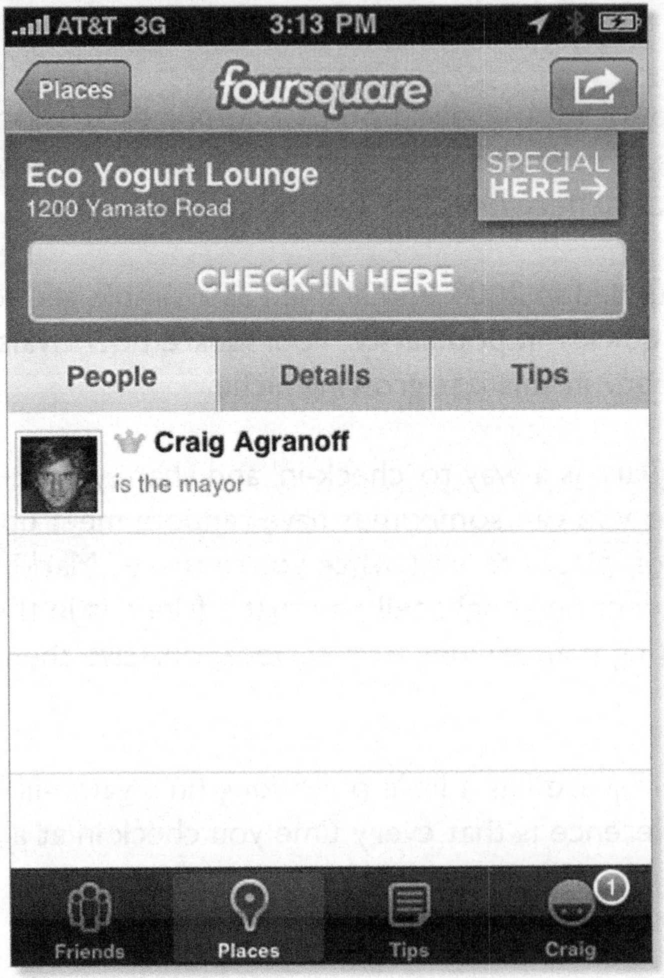
AT&T 3G
3:13 PM
Places
foursquare
Eco Yogurt Lounge
1200 Yamato Road
SPECIAL HERE →
CHECK-IN HERE
People
Details
Tips
Craig Agranoff
is the mayor
1
Friends
Places
Tips
Craig

your back yard or a stadium, you earn virtual points. The points you earn then add up for each location leaving the person at that location with the most number of points, being declared "Mayor" of the spot.

In addition to being declared the top dog, you can also earn badges for both checking into a location and for random events that might be happening there. You could be declared "Master of Ceremonies" or "Coffeehound" for visiting a location and checking in.

New additions to Foursquare are beginning to make these badges and virtual points actually pay off in reality. We will talk more about that later on.

What Are Foursquare and Gowalla Good For?

Connecting Virtual Friends in 'Reality'

While applications like Twitter and MySpace have pulled friends together "virtually" for years, they are limited in how they allow those connections to flourish in physical life. Most users of those services have the majority of their contacts (or "friends") lists made up of people they know off the Internet. Others have friends on a social networking service that they have never physically met or even talked to voice-to-voice.

This means that "online friends" usually remain only "virtual friends" in cyberspace, with no real-world connection. Until now.

With the introduction of Gowalla and Foursquare, "virtual connections" can be made into "real-life" connections as well. A friend in Milwaukee might be visiting your city and not even know that is where you live. As chance would have it, however, that friend is sitting down at a restaurant in town and updates themselves by checking in through Gowalla. Amazingly, you're just outside the same restaurant and get the notice that your friend is inside. Walking in, you check in as well and meet your "virtual" friend for the first time, face-to-face, over lunch.

This is just one of many examples of how "virtual friends" can finally meet in reality. Current friends can also randomly meet up in this same way, which facilitates all kinds of new ways to network and keep in touch.

As Fun and Friendly Competition

The game play aspects of Gowalla and Foursquare are another part of their appeal. In a fun way, friends can compete ("virtually") for titles, prestige, and bragging rights. After all, if you have your backyard swimming pool turned into a virtual spot (called tagging a geo-location) in Foursquare and your friend decides to make it "his," he can visit it often by "checking-in" and attempt to outrank you to become "Mayor" of Bob's Pool.

Similarly, you can begin a competition with your spouse to see who can collect the most Brown Shoes in Gowalla or gain the coveted "Denver Championship Chase" badge. All you have to do is issue the challenge and take the Tour through Gowalla. Whoever finishes first, wins!

There are almost limitless ways you can use these applications to enhance relationships and have fun doing it.

Building Your Business

These geo-location applications can also be used to help build your business. This is especially true of brick-and-mortar, local businesses who want to utilize the networking possibilities that these "on the street" mobile apps offer.

From giving free haircuts to the "Mayor" of your salon, to issuing virtual coupons to those who check in at your bakery, the real possibilities for enhancing a local business' customer base through geo-location are only just beginning to be recognized. The sky is almost literally the limit with this new frontier.

We'll get more into this later in the book.

Chapter 2

Getting Started With Foursquare

Foursquare is a lot of fun, but most of that fun is in the use of the app, which requires you to sign up. Unlike other social networking websites, you cannot do much with Foursquare without signing up. Twitter, for instance, allows you to read and browse streams of content without having a Twitter account, but Foursquare is very limited in that regard because, at its heart, Foursquare is an interactive social experience and that interaction requires you to be in on the conversation.

Signing up and downloading the app for your phone is free, and takes only a few minutes. In fact, you can skip typing information into the signup boxes and just directly link with your Facebook account and get going almost immediately.

Throughout this chapter, we will be looking at how to sign up, get started, and

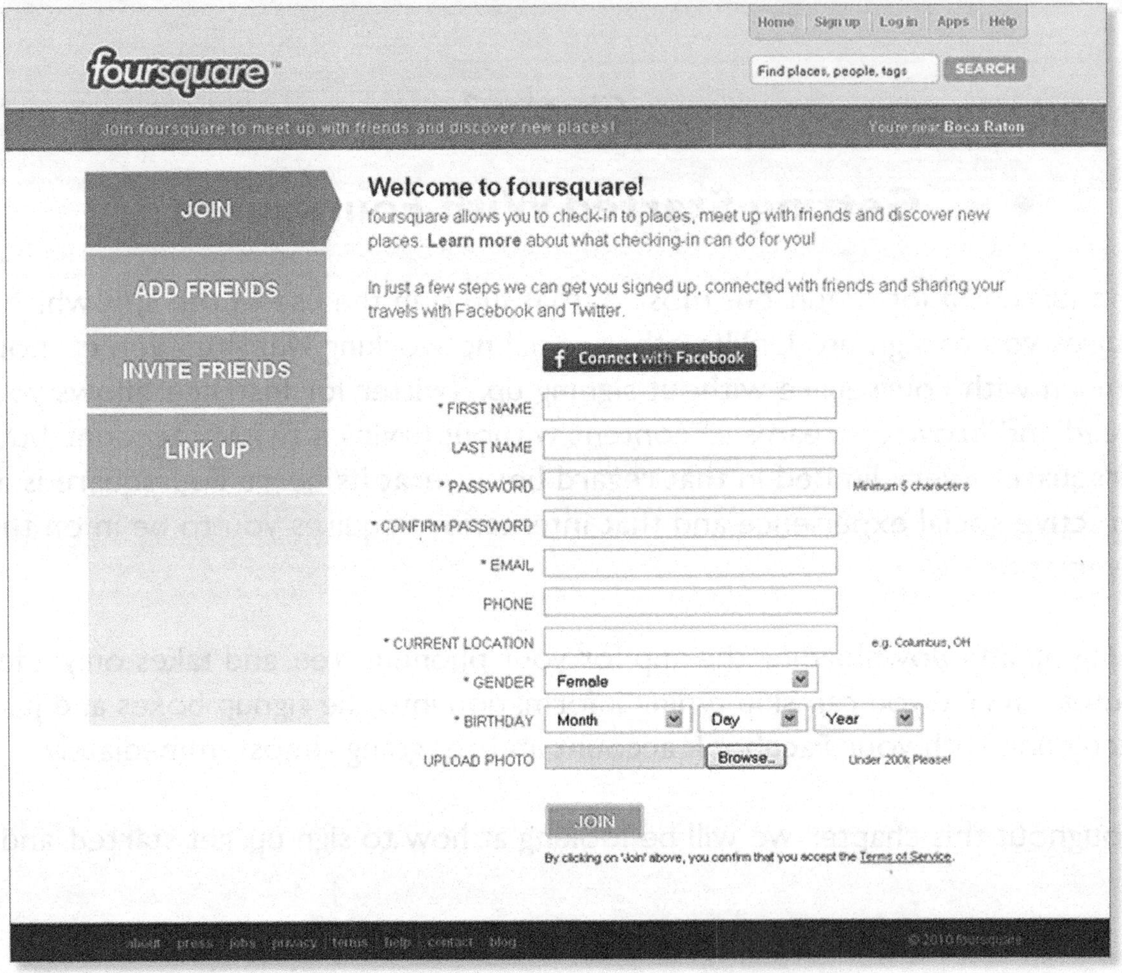
Home
Sign up
Log in
Apps
Help
foursquare
Find places, people, tags
SEARCH
Join foursquare to meet up with friends and discover new places!
You're near Boca Raton
JOIN
ADD FRIENDS
INVITE FRIENDS
LINK UP
Welcome to foursquare!
foursquare allows you to check-in to places, meet up with friends and discover new places. Learn more about what checking-in can do for you!
In just a few steps we can get you signed up, connected with friends and sharing your travels with Facebook and Twitter.
Connect with Facebook
* FIRST NAME
LAST NAME
* PASSWORD
Minimum 5 characters
* CONFIRM PASSWORD
* EMAIL
PHONE
* CURRENT LOCATION
e.g. Columbus, OH
* GENDER
Female
* BIRTHDAY
Month
Day
Year
UPLOAD PHOTO
Browse...
Under 200k Please!
JOIN
By clicking on 'Join' above, you confirm that you accept the Terms of Service.
about
press
jobs
privacy
terms
help
contact
blog
© 2010 foursquare

begin enjoying FourSquare on both the Web and on your smart phone. In the next chapter, we'll look at signing up with Gowalla and doing the same. Then we'll get more in-depth with both services and show you how to master them quickly.

Signing Up With Foursquare

The Foursquare signup screen is linked right on the front page of their website. If you're using the app on your phone, it's recommended that you sign up through your computer's Internet connection first, so you won't have to do so much typing from your phone.

On the initial signup page, you're given a series of boxes to fill in. Or you can click on the "Connect with Facebook" button and have some of the information filled in automatically for you. You'll be asked to authorize Foursquare to access your information on Facebook. When the box closes, you'll have some of the required information, such as your full name, filled in.

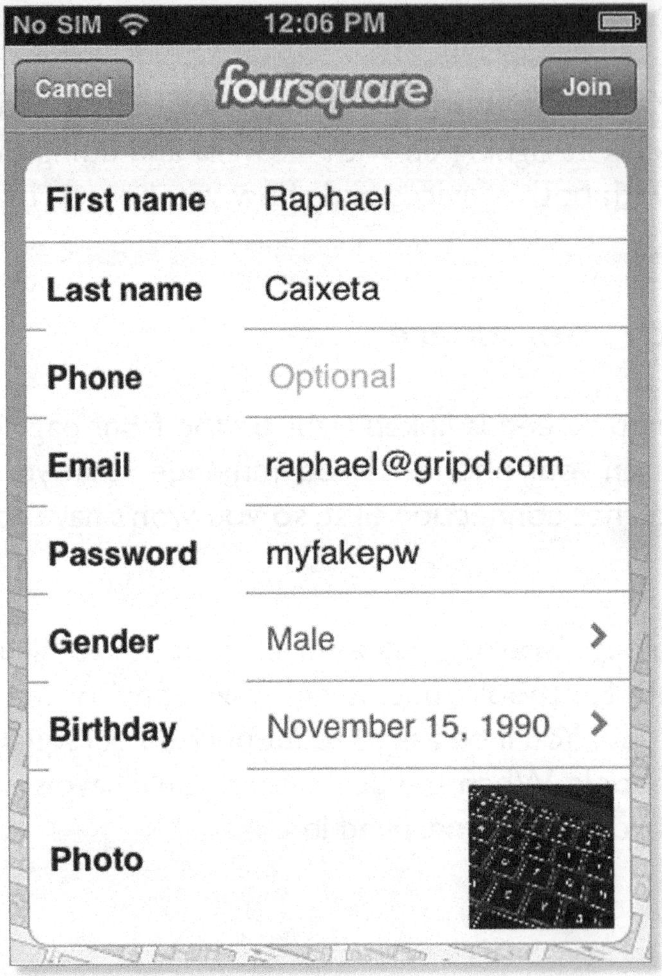
No SIM
12:06 PM
Cancel
foursquare
Join
First name
Raphael
Last name
Caixeta
Phone
Optional
Email
raphael@gripd.com
Password
myfakepw
Gender
Male
Birthday
November 15, 1990
Photo

From there, choose a solid password and, most importantly, your username. Your first name is your user name, so edit this as you'd like. You'll be known as your first name plus your location, so "Joe in Boca Raton" might be your full moniker when you're done.

Finally, choose a photo to represent you, whether it's an icon, your face, or something similar. The file needs to be small. Size the photo to about 200x200 pixels or as close to that as you can and you'll be able to upload it. Most people use their face, but anything you'd like, such as your dog, your car, or something that represents you can be used.

Finding Your Friends

After signup, two things will happen. You'll be taken to the "Add Friends" section of the signup process and you'll receive an email from Foursquare asking you to verify your email address and with links to download the Android, Blackberry, iPhone and Palm apps for your phone. Click on the verification link and keep the email so you know where to go to get the app for your phone later.

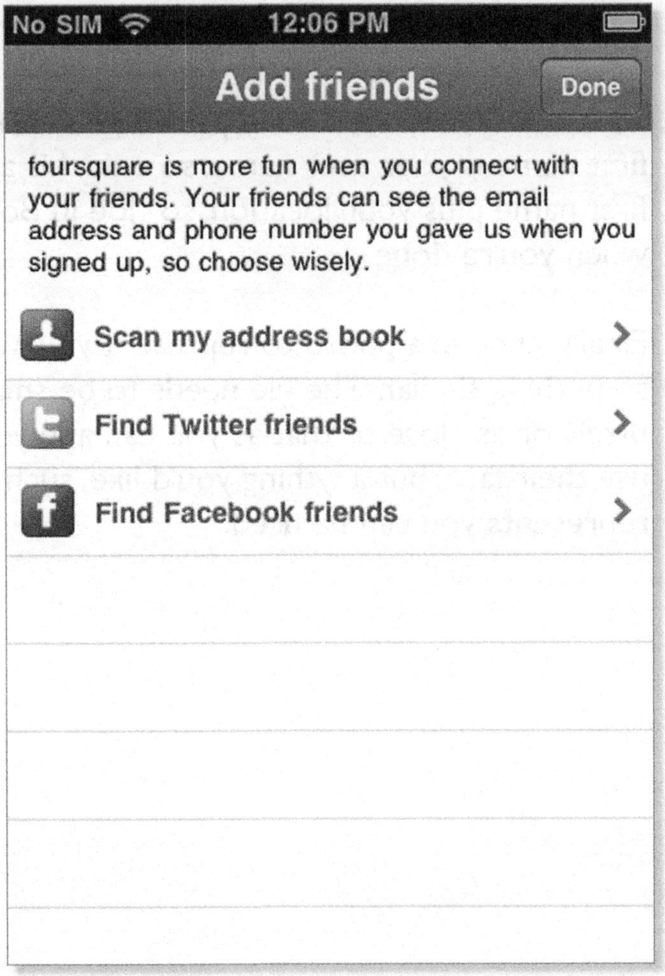

Now you're ready to begin finding your friends. If you linked with Facebook earlier, you will be able to import your friends list from there and add any of your friends who are using the app as friends in Foursquare too. You can do the same through your Twitter account. This should get you off to a very good start in synching your social networks together to make Foursquare more fun.

Before you add all of your friends, however, read the Tips for Adding Friends section later on in this chapter.

Setting Up Foursquare on Your Phone

Now that you've signed up, it's time to get your phone connected. The last page after signup and linking with your friends is the Link Up or overview page. This page has links to get the iPhone, Android, Blackberry, and Palm app for your phone. You can also set the browser on any mobile device to the mobile Foursquare page (http://m.foursquare.com/).

To use text message (SMS) connectivity with Foursquare, in the United States, you can also check in from your mobile by sending text messages to 50500. We'll discuss that more later.

For the final step in setting up Foursquare, you will want to click on the "To find your friends, click here" link. This takes you to a page that includes Facebook, Twitter and now Gmail contacts list imports. If you use Gmail, you can import your contacts from there and see if any match up with current Foursquare users. This will probably finish fleshing out your friends list for you. Again, see the next section for tips on adding friends to maximize your Foursquare experience.

Tips for Adding Friends to Foursquare

Unlike other social networking apps like Facebook or Twitter, Foursquare is not a "more is merrier" network. That is for one important reason: it's a geo-location, real-world based application. So your friends can see where you are when you're using Foursquare.

That brings up some obvious security and safety concerns, so adopting a Facebook-style "invite everyone, friend them all" philosophy might not be a good idea. As Foursquare's co-founder Harry Heyman asks, "Do you really want them to know where you are all the time?"

Probably not. So choose carefully when you add friends. Link up with those you already know in person, or know very well online. Friends, colleagues, and family are your best choices. The number of friends you have doesn't show up in your Foursquare stats list because it's really not important, so there's no competition to have a lot of them. It is important to have friends who are both active and who enjoy the same things you do. So pick wisely and screen out likely stalkers!

Building a Great Foursquare Profile

Your profile on Foursquare is how other people will find you and will be how they relate to you in the game. If you're using Twitter, consider sending your important Foursquare updates to your Twitter account automatically as tweets. To get the most out of Foursquare without being too annoying in Twitter, it's recommended that you only update Twitter with important happenings (becoming mayor, getting a badge).

Once you've linked your Twitter account with Foursquare, you'll also be able to send updates as tweets as a check box option when you check into new places or receive new things. This can boost the interactivity of the game and also bring some of your Twitter friends to your Foursquare experience.

Once again, make sure you have a picture on your profile as well. Many rewards, including being named mayor of a location, require that you have a profile picture.

Last, but not least, just using Foursquare in your everyday life can lead to new connections you didn't even know existed. Amanda, the secretary in the front office uses it and you might not have known if you hadn't seen that she's Mayor of the office building now!

Understanding Foursquare Lingo

There are several phrases or terms you might hear people use when talking about Foursquare that you may not understand. That's OK, like any new thing, there is a little learning curve. Up to this point in this book, we've used several terms you may not understand. Let's look at the common Foursquare jargon.

Check-in: this is a term used to describe checking into a new location or telling Foursquare that you have arrived in a location, such as a restaurant, your office, the club, etc. This alerts your friends on Foursquare that you have arrived in a location.

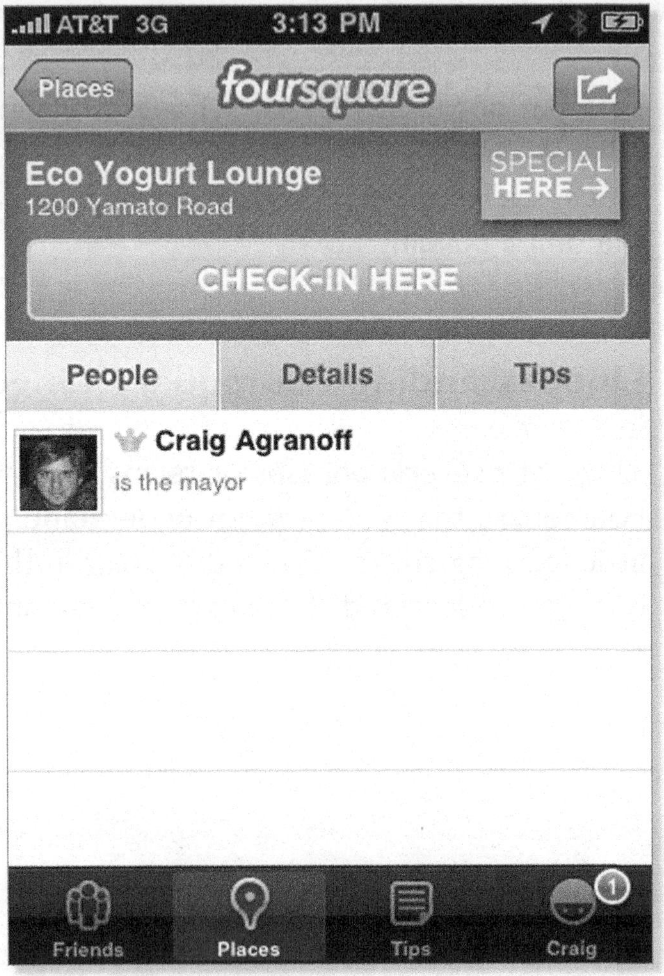

Badges: these are awards you earn for checking into interesting places either at specific times or frequently or even just randomly. When someone creates a new location (called "discovering"), they can also have the option of creating a badge to fit the venue as a "tag."

Mayor: this is the ultimate award and signifies that you have checked into a specific spot more often than anyone else over the past two months. Mayors can receive more than just the title. We'll talk more about that later.

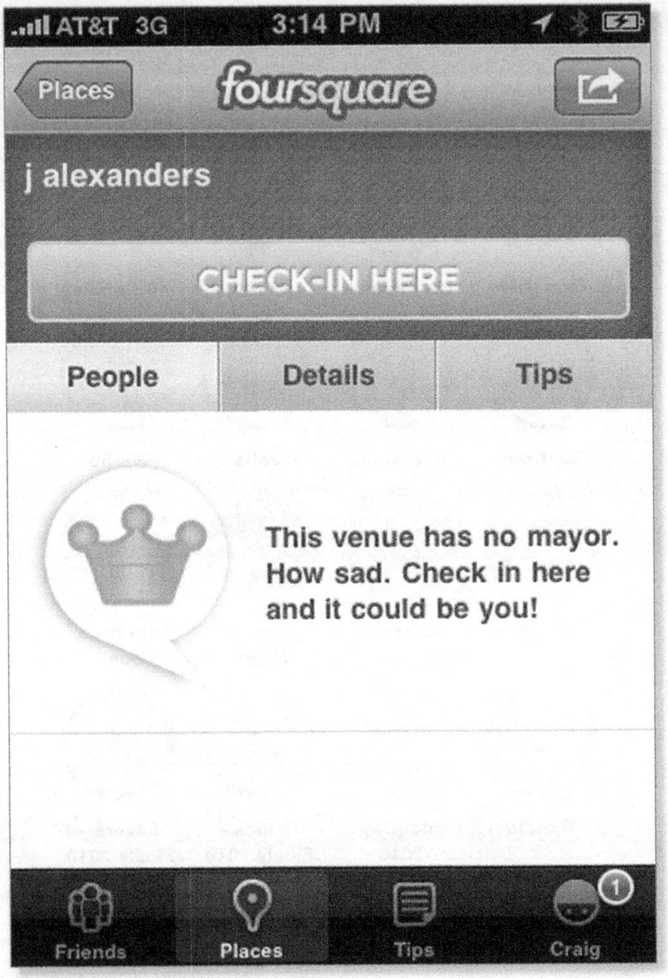

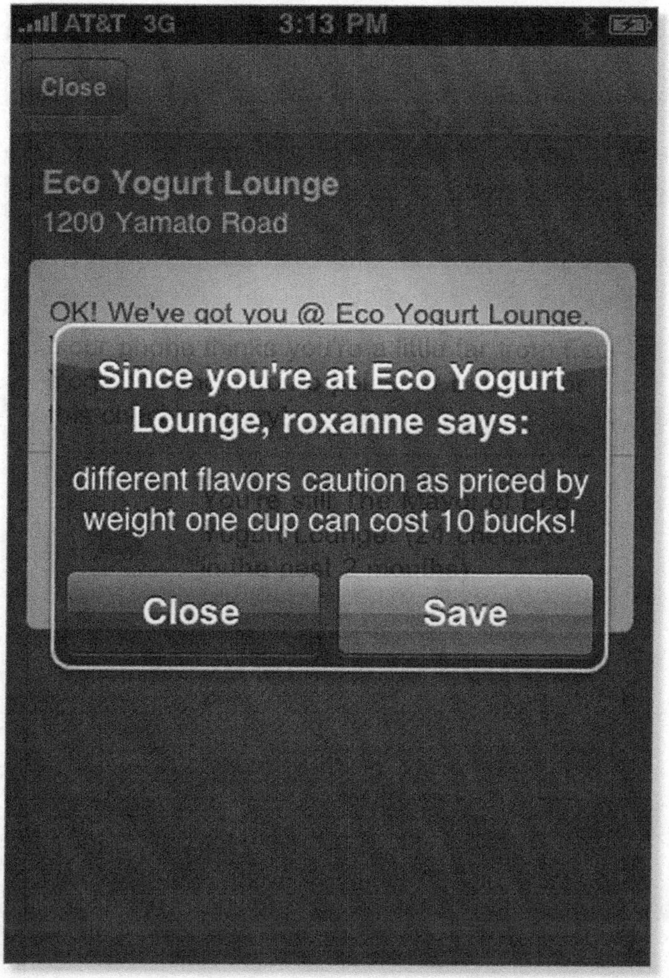

Tips / To-Dos: these are user-created tips and things to do that can be attached to places visited. They can be tips like "Get the Double Whammy Burger – totally awesome" or "Ride the Spoffaw Coaster it rocks." To-dos are tips that,when presented to you, you choose to try and put on your list of things to do.

Off the Grid: this means two things on Foursquare. Most often, it means checking into a place without broadcasting your location, which is always an option. It can also mean logging off of Foursquare (shutting down the app on your phone).

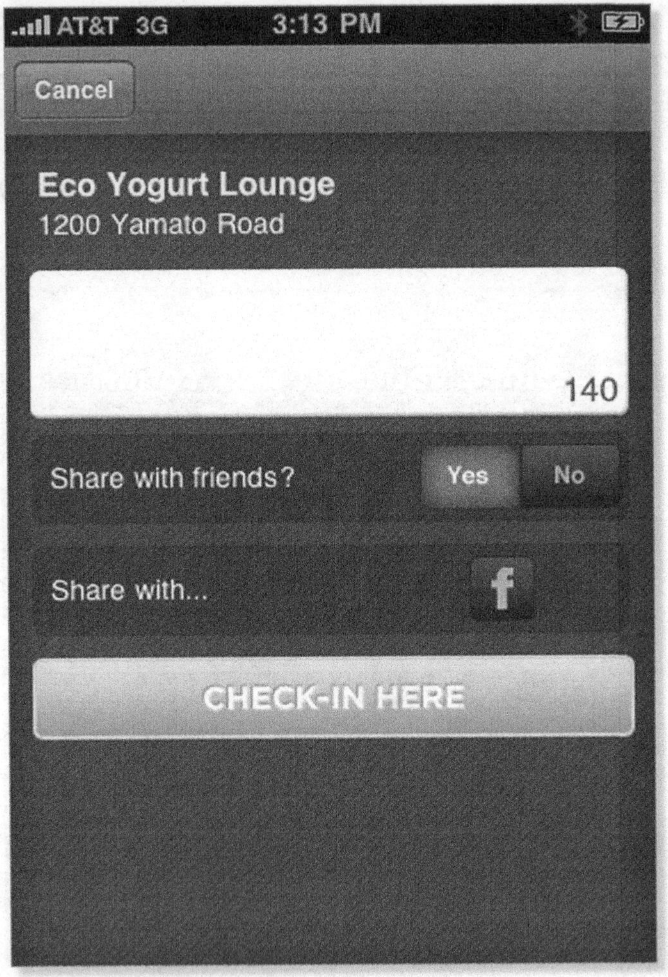

Chapter 3

Getting Started With Gowalla

Like Foursquare, Gowalla requires that you be signed in (have an account) to use the application to its fullest. Without a login, Gowalla falls flat because it is, at its heart, an interactive offline service that happens to use the Internet to facilitate those interactions. If you're not in the conversation, then you're missing the experience that is Gowalla.

The app is free to download for your phone and works on the Apple iPhone, any Android phone, Blackberry phones, and Palm operating devices. It's easier to sign up via the Web and download the app to your phone later, though.

So we'll begin this chapter by looking at the signup process (via the Web), how to get the app for your phone, getting started with Gowalla, and enjoying the app. In later chapters, we'll look at more in-depth and day-to-day usage of both Foursquare and Gowalla and more.

Signing Up With Gowalla

Going directly to Gowalla.com, you are presented with a large welcome screen that explains the application, gives you links for downloading it, and displays a large signup button to the right. That is where we'll start. Signing up through your computer's Internet connection is usually easier than using your smart phone.

Clicking the signup button, a very simple and easy form is presented. It only has five boxes to fill in. Most are self-explanatory, but let's look at the two most important ones: your username and your password.

The password you choose should be solid and secure, just like most of your passwords should be. The username is how you'll be known (or viewed) on Gowalla and is where your future Gowalla URL will be pointed as well.

Choose a username that reflects who you are or what your business is about, depending on what you're using Gowalla to accomplish. Names do better than

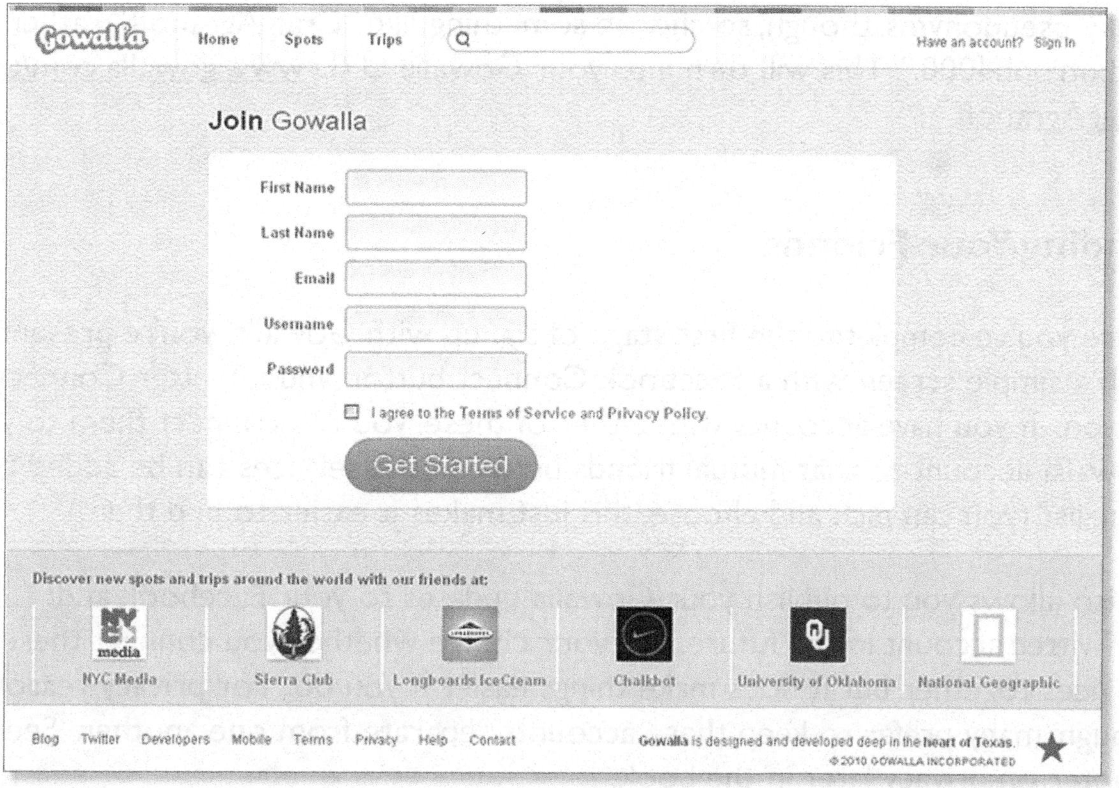
Gowalla
Home
Spots
Trips
Have an account? Sign In
Join Gowalla
First Name
Last Name
Email
Username
Password
I agree to the Terms of Service and Privacy Policy
Get Started
Discover new spots and trips around the world with our friends at:
NYC Media
Sierra Club
Longboards IceCream
Chalkbot
University of Oklahoma
National Geographic
Blog
Twitter
Developers
Mobile
Terms
Privacy
Help
Contact
Gowalla is designed and developed deep in the heart of Texas.
© 2010 GOWALLA INCORPORATED

geeky pseudonyms, though, so choose something like "CraigAgranoff" rather than "Maxitronn4000." This will turn into your Gowalla URL: www.gowalla.com/users/CraigAgranoff

Finding Your Friends

Once you've completed the first stage of signup with Gowalla, you're presented with a simple screen with a Facebook Connect button and a Twitter Connect button. If you have accounts with either of these, you can connect them to your Gowalla account so that mutual friends between the services can be added to your list (you can pick and choose, this just makes it easier to find them).

It also allows you to publish your Gowalla updates to your Facebook and/or Twitter account in the future. It's your choice whether you connect these services together, but it does make things easier if you do. For privacy reasons, though, many prefer to keep these accounts separate from one another. See our chapter on privacy later in the book.

No SIM
11:35 AM
Gowalla
Phone
Mail
Safari
iPod

No SIM
11:35 AM
Sign In
Sign Up
Save
Gowalla
Autofill using Address Book
First Name First
Last Name Last
Email me@example.com
Username username
Password 3 characters required

No SIM
11:36 AM
Sign In
Sign Up
Save
Gowalla
Autofill using Address Book
First Name Raphael
Last Name Caixeta
Email raphael@gripd.com
Username raphaelcaixeta.
Password ●●●●●●●●●●●

By now, you should also have received an email from Gowalla welcoming you to the service. That email contains easy links to download the smart phone apps, so you should keep it handy for our next section. After that, we'll discuss adding more friends to your Gowalla network. For now, leave your browser open where it is and let's look at your phone for a minute.

Setting Up Gowalla on Your Phone

Now that you have a login for Gowalla, let's get your phone connected too. There are two ways to do this, depending on your phone's configuration, the first and easiest way is to either forward the welcome email Gowalla sent you with the links to your phone or simply access that email from your phone. In either case, the goal is to get the links to your phone easily.

The second option is to visit the application directly. For the iPhone, go to the AppStore to download Gowalla, for the Android you'll need to visit the Android

Market, for Palm visit the Palm Catalog, and for Blackberry, point your browser directly to gowalla.com.

Once you've installed the app, it will ask for your login information. This is the username and password you used to initially sign up. Enter that information, and you're ready to roll with your phone. Now let's look at adding more friends to the service.

Tips for Adding Friends to Gowalla

After the initial Facebook/Twitter link page, you are taken to a page with several more services you can link through. These include Gmail, Windows Live, and others. If you have a login for any (or all) of those, you can use that as you did the Facebook/Twitter links to synch the services and find friends that are also using Gowalla.

It's important to note, however, that unlike Facebook or MySpace, Gowalla is not as well-suited to the "more is merrier" philosophy of networking. Instead, it's a service based on real-world contacts or at least geographically similar contacts. So adding all of your worldwide friends may not be conducive to a fun experience on Gowalla and may isntead overwhelm your stream with updates from people you don't really know and never have contact with.

The other question to ask yourself is whether you really want whomever you're going to connect your Gowalla account with to know where you are at all times? To begin with, it's best to stick with personal friends, family and colleagues you know well. You can always add more people later.

Building a Great Gowalla Profile

After you've connected with some friends, you will need to build your profile on Gowalla. This is how other Gowalla users will view you and what will be

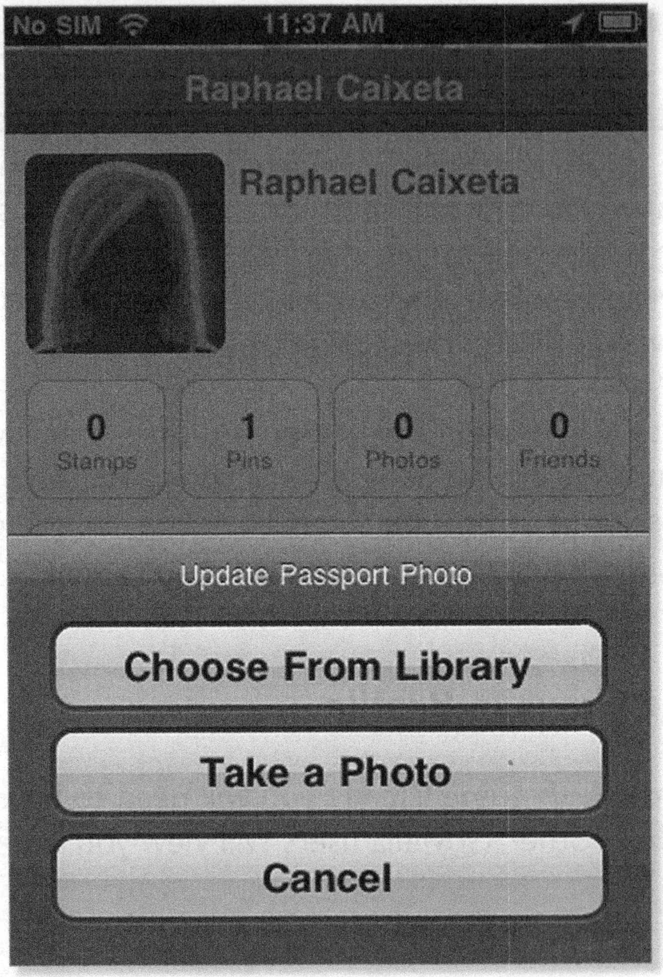
No SIM
11:37 AM
Raphael Caixeta
Raphael Caixeta
0 Stamps
1 Pins
0 Photos
0 Friends
Update Passport Photo
Choose From Library
Take a Photo
Cancel

presented on your Gowalla page when people navigate to your profile. Fill in as much information as you'd like, but remember that everything except your email address and telephone are shown publicly.

Choose your email preferences and then a photo. Select a good profile picture. Pictures of yourself (usually your face) are most popular, but anything that both conveys who you are and fits with your plans for Gowalla is a good choice. These will always be thumbnails, so try to keep the photo small (250x250 pixels) if you can.

Your first and last name will be displayed, so edit these to match your personal privacy preferences. Most people, if they're searching profiles, will look for specific names and locations, so be sure that enough information is there that your friends will be able to both find and recognize you. A first name, your last initial, and your city and state will probably do the trick, but if you don't mind having your full name there, leave it that way.

Fill in the About area with a short sentence or two describing who you are (your job, your hobbies, or similar). You don't need to be too specific ("Marketing" is better than "Marketing for XYZ Co"), but keep in mind your planned uses for the account (business or pleasure or both).

Finally, click on the Privacy tab and turn your passport privacy to "On" so that only your friends can see your location, updates, etc. You can change this later if you wish, but it's recommended that you begin with it activated for security reasons.

Understanding Gowalla Lingo

There are a lot of phrases and terms used by Gowalla fans that may not make sense to you without some explanation. To help you with that, here is a short list of some of the most common ones.

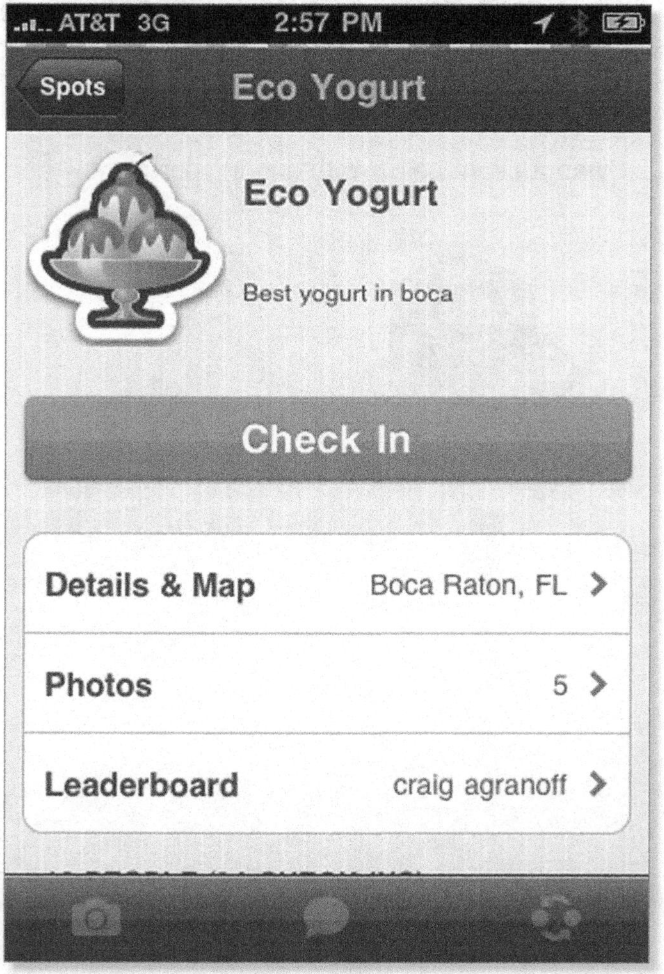

Check-in: as with Foursquare, this is the act of announcing your arrival at a location. You are both telling Gowalla to record you as being there (for game purposes) and alerting your friends on the network as to where you're currently located.

Drop: these are items that have been "dropped" in locations for other Gowalla users to find. Some are random from the app itself, but anyone who visits a spot and checks in has the option of leaving behind something for someone else to find. These items contain a history of the past owners and where they've been and are part of the fun of Gowalla.

Passport: your itinerary of where you've been and what you've accomplished on Gowalla. It is what is displayed at your Gowalla URL (www.gowalla.com/users/yourusername).

Spots: these are locations in Gowalla that have been created or that you are creating. So if you attempt to check in at a location that does not exist on Gowalla yet, you are given the option of creating that spot. Once you do, your name (username) will be linked to it forever.

Trips: are tours of an area that are taken by Gowalla users. They are a series of spots all combined to make a connected tour. Trips are a fun way to use the app and can become competition as players attempt to make the trip faster than their friends, earning badges and awards for doing so.

Off the Grid: is also the same as Foursquare and merely means you've either switched off your phone or you've closed your Gowalla app so that you are no longer active in the game. Many users will log off of Gowalla by announcing they're "going OG".

Chapter 4

Foursquare on Your Phone, Everyday

Using Foursquare every day is how the game is played. Of course, it's not a total life-sucker like some online games. It's more a kind of simple daily thing you do to keep connected with your friends while doing most of the things you normally do anyway.

The interactions you can create with Foursquare are not time-consuming, but they can be both fun and relationship-building. Small competitions with your friends, an afternoon hitting a few hot-spots to raise your scores, or a spontaneous meetup for lunch are the things that make the Foursquare app what it is.

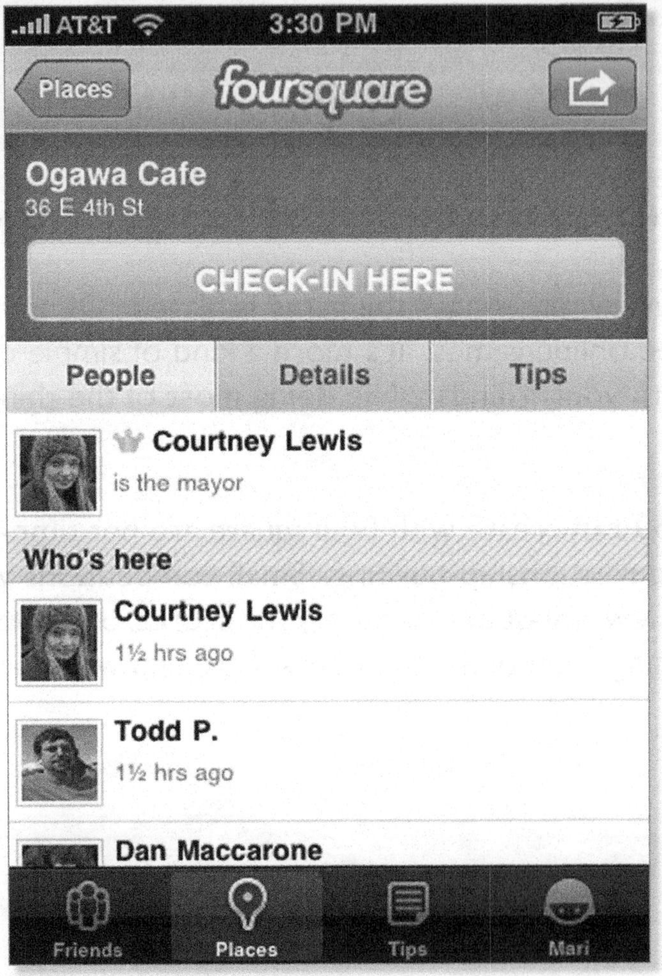
AT&T
3:30 PM
Places
foursquare
Ogawa Cafe
36 E 4th St
CHECK-IN HERE
People
Details
Tips
Courtney Lewis
is the mayor
Who's here
Courtney Lewis
1½ hrs ago
Todd P.
1½ hrs ago
Dan Maccarone
Friends
Places
Tips
Mari

Checking In

Everywhere you go is a possible check-in opportunity. Of course, checking into the bathroom at the local bus terminal might not be exciting, but your morning visit to get coffee, your lunch at Tucci's Pizzeria, and the quick shopping stop at the grocery on the way home are all possible check-in points. Just because a place seems mundane doesn't mean it is and "hot spots" for competition in Foursquare can happen anywhere, even at a bus stop.

Using the Foursquare app, check-in at whatever location you'd like, so long as it's a fixed and publicly-accessible spot. The mailbox in front of your house is fair game (anyone can walk by there), as is your favorite restaurant or even the gym.

If a location doesn't exist yet, feel free to add it. Look ahead to the 'Adding a Location' section for more information.

Making Your Check-ins More Memorable

Checking into a location can be as simple as tapping the app and acknowledging that you've arrived in a location. That, at its essence, is what check-ins are all about. It's boring.

Try adding shout outs (messages) to your check-ins to engage friends who might frequent that spot. Or invite someone who hasn't been there before to try it out. Or just rub it in that you now have more check-ins there than your friend does.

Even just throwing out a public service message telling your friends what's happening at your current location is better than just checking in. Be creative and fun with your check-ins.

Leaving Tips At Locations

No, not tips to the waiter, these are tips to your friends or other visitors to the location. Tips are a fun way to give more information about a place and, for the business owner, a chance to draw in customers from the street.

Tips are sort of like crowd-sourcing (via Foursquare) for the location you're in. Like Yelp or other services that "rate" places based on what users have left on their site for feedback, leaving Tips at a Foursquare check-in allows you to not only tell people how good (or not so good) a place is, but also lets you recommend things or give people tasks to try while they're there.

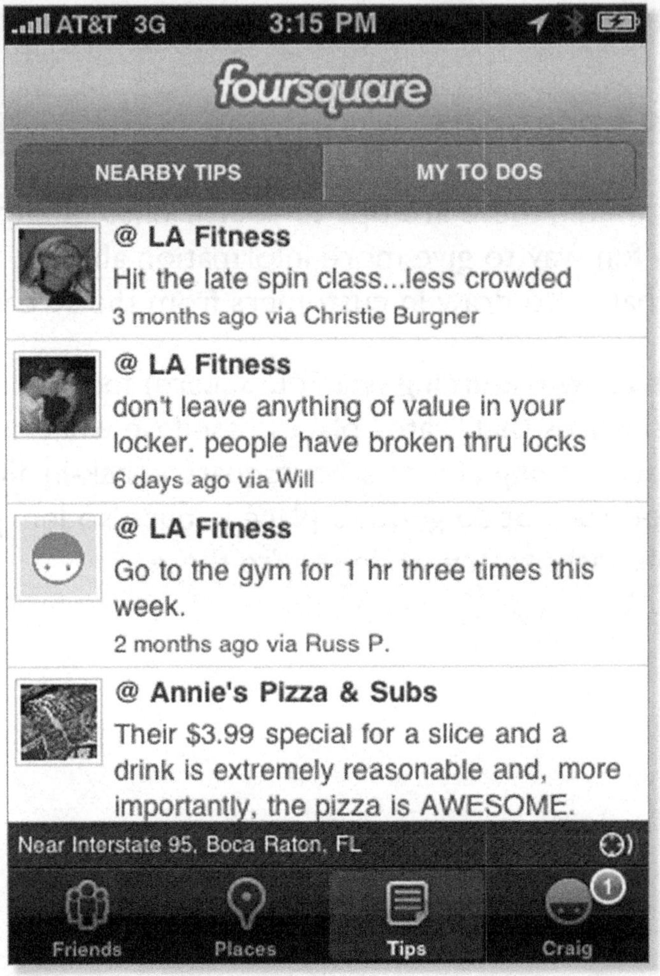
AT&T 3G
3:15 PM
foursquare
NEARBY TIPS
MY TO DOS
@ LA Fitness
Hit the late spin class...less crowded
3 months ago via Christie Burgner
@ LA Fitness
don't leave anything of value in your locker. people have broken thru locks
6 days ago via Will
@ LA Fitness
Go to the gym for 1 hr three times this week.
2 months ago via Russ P.
@ Annie's Pizza & Subs
Their $3.99 special for a slice and a drink is extremely reasonable and, more importantly, the pizza is AWESOME.
Near Interstate 95, Boca Raton, FL
Friends
Places
Tips
Craig
1

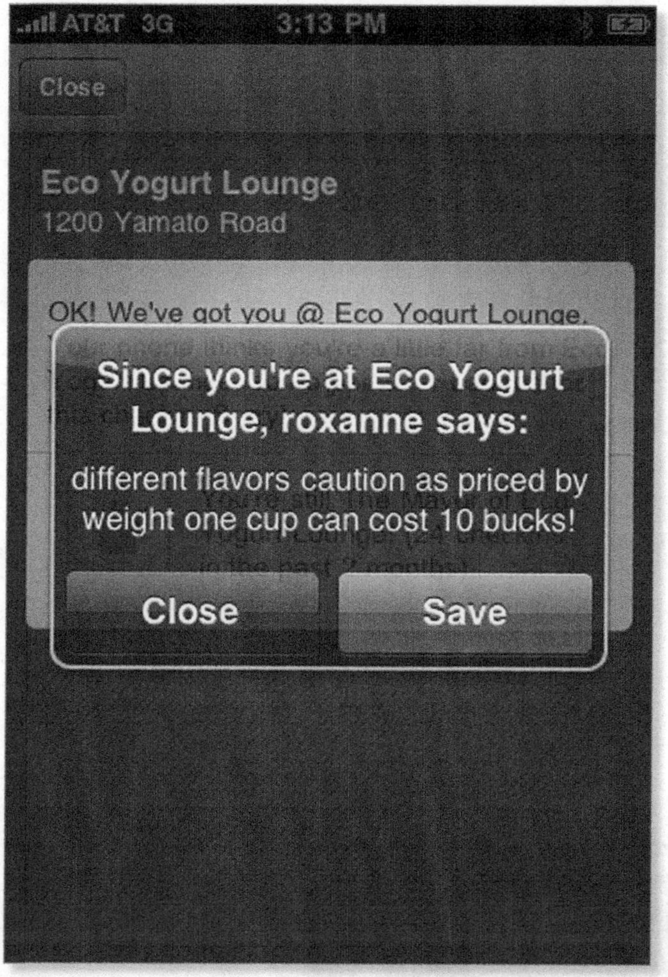

You can leave little blurbs about a restaurant, for instance, telling people that the cheese pizza is the bomb or the guacamole burger is the best in the city. These will appear as lists whenever someone is near the location with their Foursquare app open. Upon checking into the location, people can see all of these recommendations and even add them to their 'To Do' list so they can try them now or on another visit.

AT&T 3G
3:29 PM
foursquare
NEARBY TIPS
MY TO DOS
@ The Place (Waffle's)
Be nice to the Mayor, and maybe he will make you a pizza.
5 months ago via Michael Katz
@ Cafe Bella Sera
Remember to try the pizza if they have some here for www.worstpizza.com
6 months ago via Craig Agranoff
Near Interstate 95, Boca Raton, FL
Friends
Places
Tips
Craig

Your To Do List and Top 12

Your 'To Do' list is a list of things you've seen as Tips at various locations and have listed as things you want to try as well as things you've manually added to your list when visiting a location. Maybe you were torn between trying the ferris wheel and the double-dipped chocolate cones at the ice cream stand at Fun Land and really only had time for one. Well, the other can become a To Do.

The Top 12 list on your To Do screen is made up of those things you'd most like to do, plan to do as an itinerary (of sorts), or otherwise wanted to emphasize.

Maybe you and your girlfriend are going out on a Foursquare Date and have created Top 12 lists to do while you're out. Sounds geeky, but try it and you'll see that it's actually fun.

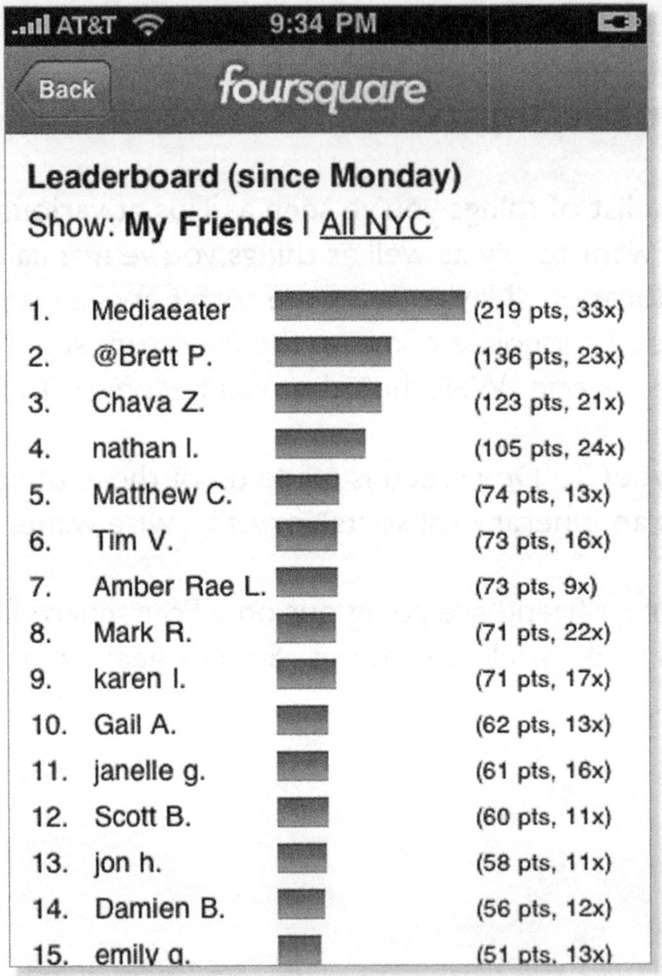
AT&T
9:34 PM
Back
foursquare
Leaderboard (since Monday)
Show: My Friends | All NYC
1. Mediaeater (219 pts, 33x)
2. @Brett P. (136 pts, 23x)
3. Chava Z. (123 pts, 21x)
4. nathan l. (105 pts, 24x)
5. Matthew C. (74 pts, 13x)
6. Tim V. (73 pts, 16x)
7. Amber Rae L. (73 pts, 9x)
8. Mark R. (71 pts, 22x)
9. karen l. (71 pts, 17x)
10. Gail A. (62 pts, 13x)
11. janelle g. (61 pts, 16x)
12. Scott B. (60 pts, 11x)
13. jon h. (58 pts, 11x)
14. Damien B. (56 pts, 12x)
15. emily g. (51 pts, 13x)

Competing for Honors

Once you've got the hang of Foursquare and are using it regularly (or hourly for some of us), you can start thinking about competition. There are a lot of them on the app and each has its own little challenges and fun. Most of them center around checking into locations.

Some honors are centered on being the one to check in most often at a location (Mayor) while others are based on the time of day you checked in (Night Owl) or the way you checked in (Socialite).

When you check into a location, click on the "unique visitors" or "checkins here" buttons to see more than just visitor counts. This will show you what titles have been won here, and by whom (if public), as well as some other information. The Mayor button shows the current mayor of this location and that button can also give you information on the number of visits in the past two months that person made to win the title.

NERD MERIT BADGES
1.5", fully embroidered, Velcro-backed. Attach to your jacket, your backpack, or the lid of your over-clocked, battle-scarred laptop. Start a nerd sash!
Lose It or Lose It: FLAWLESS VICTORY
This badge can't be purchased, only earned!
In addition to making badges, both Randy and...
$0.00
Add to Cart
Laptop Sash
A seventeen-inch length of two-inch wide, self-adhesive, loop-side, low-profile Velcro. The Laptop Sash sticks permanently...
$13.37
Add to Cart
Foursquare Badge: Douchebadge
"Double pop that collar, son!"
Earned for checking in multiple times to venues tagged "
$5.99
Add to Cart
Foursquare Badge: Gym Rat
"10x trips in 30 days? Go reward yourself to the nastiest double cheeseburger you can...
$5.99
Add to Cart

There are other, hidden titles and badges of honor you can win throughout Foursquare while visiting locations, so check in everywhere you can!

Note: if you want the badge in real life, you can buy it from Nerdmeritbadges.com, which has officially been given license to embroider and sell Foursquare badges.

Adding a Location to Foursquare

Adding a location (venue) to Foursquare is easy and fun. If you own a business, run a non-profit, or frequent a place that you'd like to shout out to your friends, you can add it to Foursquare. Just click the 'Add Venue' menu item and start filling in information.

Let's say you're visiting the local animal shelter and want to add them to Foursquare so you can check-in whenever you come again. First, be sure they aren't already on Foursquare by attempting to check in. If it doesn't know where you are, select "Add Venue" and enter the name of the organization.

Now put in the address, abbreviating street names ("street" becomes "St.") and spelling out the city, state (all caps), and ZIP (if known). Country codes should be in all caps as well. If the organization has a Twitter ID, include that as well, otherwise skip it.

Add some tags to make it easier to search for your venue (hint: include the city/state as a tag).

You're done! Now that it's been added, why not throw in some tips or send a shout out to your friends about the new location? Who knows, in two months, you could become Mayor of the Animal Shelter!

Going Stealth

The term "going stealth" literally means you're going off the grid with your Foursquare check-ins so that nothing is broadcast to your network of friends. You can do this for a variety of reasons: maybe you don't want your boss to know you're not back from lunch yet, or maybe you really wanted to just sit in the restaurant to read a book over lunch and don't want your friends to know.

Whatever your reasons for Going Stealth, it's easy to do. Just choose the "Off the Grid" option when checking into a location and you'll be in stealth mode.

Now pull the hat down over your eyes, check your sunglasses, and make sure that false beard is taped on straight and you're good to go.

Chapter 5

Gowalla on Your Phone, Everyday

Unlike Foursquare, Gowalla is not as centered on how often you check in so much as it is about the items you can get or leave behind while you're there. For the more casual user who doesn't want to compete based on how many times they can score points, Gowalla is the more laid-back choice.

Another difference between the two apps is that in order to check into a location with Gowalla, you have to physically be there (well, your phone does, anyway). While for most users, Gowalla is a daily thing that is just another fun thing you do as your day progresses, of the two apps, Gowalla is definitely the one that is more real-world based.

The interactions in Gowalla are real, fun, and involve a mashup of actually interacting in the real world and leaving surprises and trading gifts in the virtual one. Whether your friends are next door or halfway around the world, you can track and interact with them in a fun way with Gowalla.

Checking In

When you arrive at a location, if your Gowalla app is already running, it will pop up with the location and give you the option to check in at that location. While checking in, you can publish your location status on Facebook, Twitter and other networks you have connected with Gowalla (your Gowalla network is automatically alerted).

Once you've checked in, you can go through your virtual backpack and leave behind something for the next person to retrieve and you'll be notified of anything that might already be in your current location that you might want to take with you.

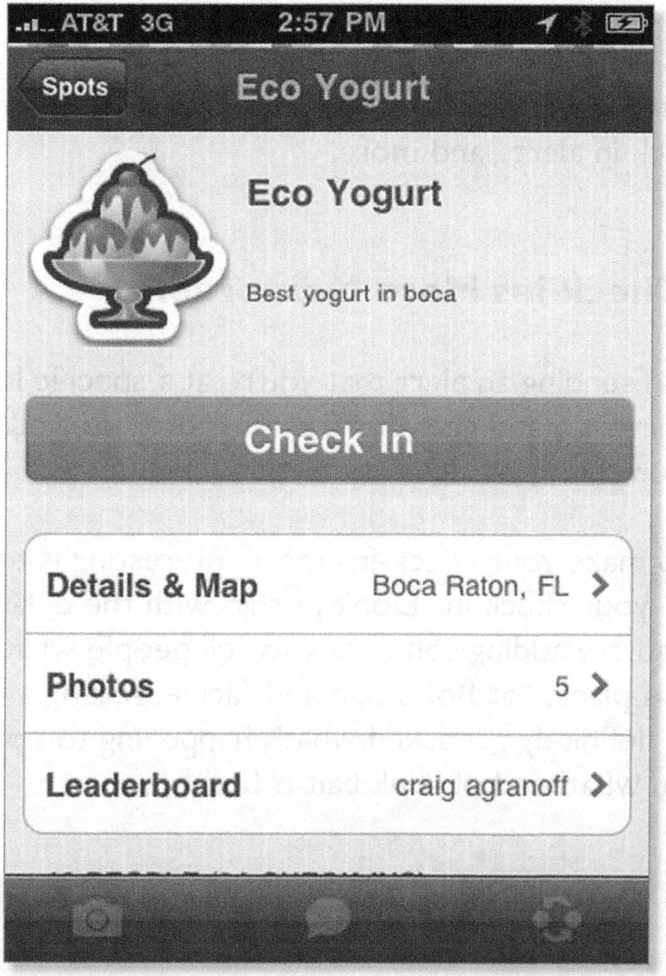
AT&T 3G
2:57 PM
Spots
Eco Yogurt
Eco Yogurt
Best yogurt in boca
Check In
Details & Map
Boca Raton, FL
Photos
5
Leaderboard
craig agranoff

The next few sections of this chapter will show you how to add locations, add pizazz to your check in alerts, and more.

Making Your Check-ins More Memorable

Just checking in and sending an alert that you're at a specific location is, well, boring. To spice things up and get your friends more involved, you should add some pizazz to your check-ins. It's easy and can be fun too.

The easiest way to make your check-ins more interesting is to add information before you publish your check in. Don't just go with the default "at Bob's Bait and Tackle." Instead, try adding something to tell people what you're there for or why you like this place. "at Bob's Bait and Tackle, getting a tub of stink bait for tomorrow." You'll definitely get asked what's happening tomorrow. Someone might even tell you what kind of stink bait is best.

You can also add photographs from your phone, so your above update and check in might also include a picture of the mentioned can of stinkbait (or you tasting it to "check it out").

Alternatively, tell your friends what's happening at Bob's or interesting items that are on sale. The possibilities are nearly endless and anything that isn't the default alert is likely to be more interesting and fun.

Dropping and Retrieving Items

The center of Gowalla's fun is in the items. Your virtual backpack is limited to ten items that you can carry at any time. This encourages you to swap out Gowalla items regularly, which makes the game more fun and interesting.

When you first create (check in) at a new spot, you can become the "founder" of that spot *if you drop an item for the next person to find.* This has several reasons

AT&T 3G
2:58 PM
Eco Yogurt
Cancel
Swap an item below with one in your pack:
Baseball Cap
#27,113
Margarita
#10,157

for being. Your "sacrifice" of giving up something without return is paid off by becoming the recognized founder of the location forever.

At most locations, items will be present and you can take them if you'd like. Every item you take, however, requires that you leave something in its place. Gowalla calls this "karma giving" and it's central to keeping items juggling and fun in the game.

Every item you pick up or that you have in your pack can be inspected to show you where it's been, who's had it, and more. Items often travel the world and that watermelon you just picked up might be like a suitcase with dozens of stickers on it from all kinds of exotic locations.

Sometimes, you'll be rewarded with random items, which cost nothing and will help to fill in your pack. It's now also possible for businesses to purchase items to give away in Gowalla – we'll get to that later.

Taking a Trip on Gowalla

Trips are one of the most popular things on Gowalla. A trip is simply several spots (locations) put into a linear order for visiting. They range from pub crawls to museum tours and more. Students at universities have created trips to make a "tour" of their university campus and hot spots. Spook enthusiasts have made trips to show and explain haunted spots through a city or area. If you can make a list of spots to visit and tie them together in a theme, it can be a Trip.

Taking a trip is easy. Just search for one on the "trips" tab of the Gowalla app and click on your chosen trip. The display of available trips in your area shows you how many spots they are in and you can click to see a map of where the trip goes so you can get an idea of how much traveling will be required.

Trips are a fun way to discover new places, plan a quick and fun date, or create some crazy competition amongst friends as you all compete to finish a trip first. Trips don't require that you do all of it at once and as you check in at locations

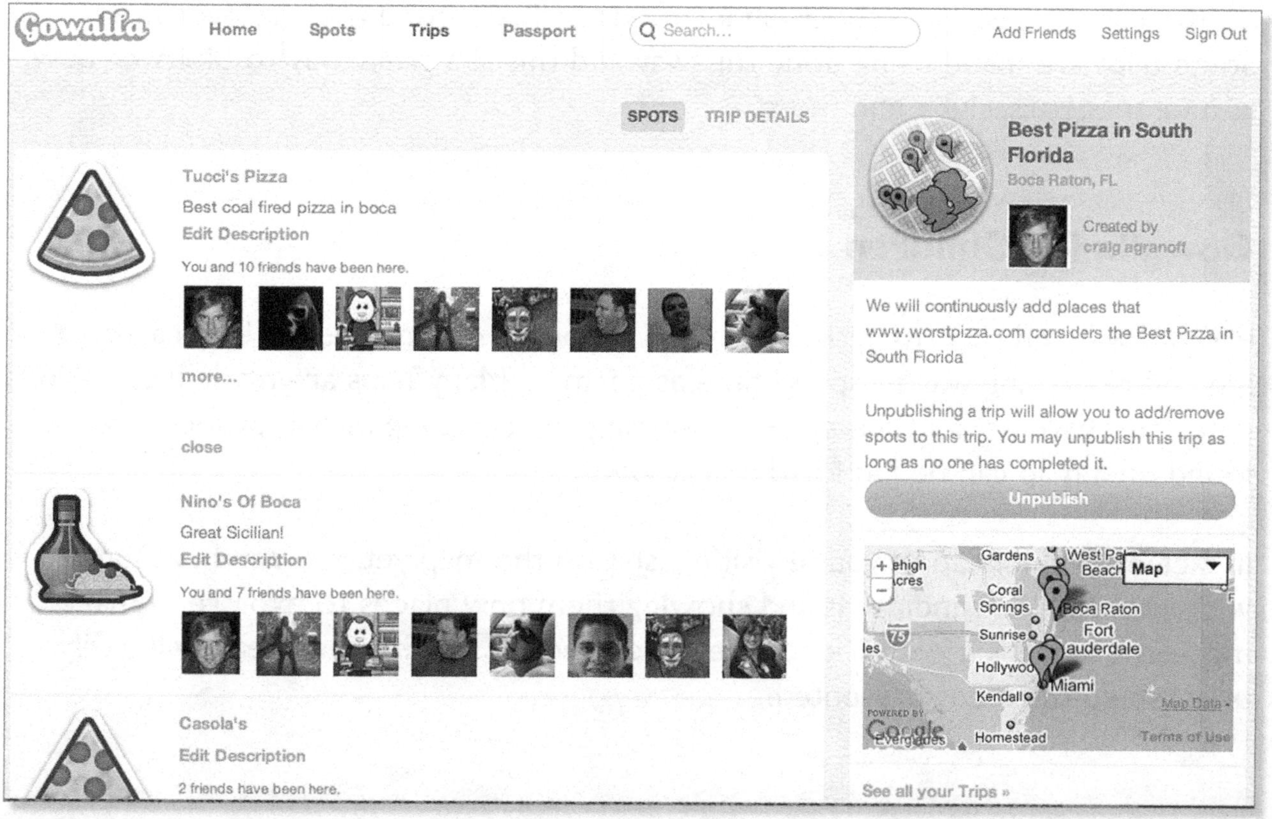
Gowalla
Home
Spots
Trips
Passport
Search...
Add Friends
Settings
Sign Out
SPOTS
TRIP DETAILS
Tucci's Pizza
Best coal fired pizza in boca
Edit Description
You and 10 friends have been here.
more...
close
Nino's Of Boca
Great Sicilian!
Edit Description
You and 7 friends have been here.
Casola's
Edit Description
2 friends have been here.
Best Pizza in South Florida
Boca Raton, FL
Created by
craig agranoff
We will continuously add places that www.worstpizza.com considers the Best Pizza in South Florida
Unpublishing a trip will allow you to add/remove spots to this trip. You may unpublish this trip as long as no one has completed it.
Unpublish
Map
Gardens
Coral Springs
Boca Raton
Sunrise
Fort Lauderdale
Hollywood
Miami
Kendall
Homestead
POWERED BY Google
Map Data
Terms of Use
See all your Trips »

on the trip map, you can check off locations and complete the trip over time. Some trips are made to be done this way and this is a great way to discover new restaurants, bars, clubs, and more.

Gowalla for Children

Parents will find that using their Gowalla accounts to amuse their kids can be a fun and rewarding way to spend time as a family. Many Trips are made just for kids (and their parents) and the thrill of "winning" by checking into a location you've found on a map can be fun for the little ones.

In fact, even if a location you're visiting isn't on the map yet, you can have fun with your kids by "finding" it and showing them new places to explore. What's more, the app is very kid-friendly because even those who can't read will easily understand the icons presented.

There can be a lot of fun with Gowalla when you use it with children to enhance the discovery and good times. Why not craft a Trip with your children of their favorite places to go? You could map a child's tour of Springfield, for instance, with spots like the zoo, parks, museums, and more!

Adding a New Spot

When you discover a new spot (try to check in and are first to do so), you can add some information about it to make it official and then leave an item and become the spot's Founder.

First, once you've clicked the check in button at a location, you should be presented with a box that says the spot is not on the map yet. You can then add a place name, some basic information, and choose an appropriate icon for the spot.

Place names should be proper names ("Joe's Bar" instead of just "joes"). The category and icon should be appropriate as well. Joe's Bar probably doesn't fit in the "family fun" category. In the description for the spot, tell what it is and other info about the place, but don't get editorial on the quality of the food or the ugliness of the spot's owner. "Italian – American food" is a better description than "Greasy meatballs and some hamburgers made by a guy who doesn't believe in bathing." Really.

All new additions are checked by Gowalla employees and editors (called the "Street Team Elite"), so if you do misspell something or forget to capitalize a letter, it will get fixed.

Once you've discovered a spot on Gowalla, be sure to drop an item there so you can become the founder and have your name listed forever on the spot's Gowalla listing. This gives you no end of Gowalla fame and glory!

Chapter 6

Geo-Location Games

Fun Ideas for Socializing on Foursquare and Gowalla

Above all else, these geo-location apps are about having fun. Foursquare has built-in competition with the points system, but that only goes so far. Gowalla's strong point is its focus on item trading and both applications have something that no other social networking game online has: a great mixture of real-life interaction with virtual contacts.

In this chapter, we're going to look at fun games and activities you can use Foursquare and Gowalla to do. These games include ways to play in real-time, over periods of time, or otherwise interact with your friends in both real life and virtually. We'll also look at fun (and safe) ways to meet new people and create social interaction offline after you've met someone online.

So let's get started!

FourGow Steal the Flag

(4+ players, 2 locations)

If you've ever played steal the flag in school or with friends, this is the virtual equivalent. It's a mixture of the steal the flag you played as a kid and the virtual steal the flag games that are played online as game apps or website trolls.

The idea is to set up two objects (physical works best, but items in Gowalla can also be used) in two locations, which will be the "home base" for two teams of players. Once the teams and items are set, the game begins.

Each team must work to get to the other team's location either directly or, for added fun, through a set route of check-ins. To secure an object, the person(s) "stealing" it must outnumber those guarding it and must return it safely to their own home base without losing their own flag. Once the two flags are combined, the game is won.

At any time, if any team's numbers in one location outnumber those of another, then the larger team has advantage. So, for example, a flag is stolen from a home base by two team members who then check in at another location en route to their own home base to secure the prize. At that check in location, they are confronted by three team members from the opposing team, who take back their flag.

The game is over when both flags are united at one or the other home base. Whoever's home base is occupied by both flags is the winning team.

Gowalla Cyberstalker

(2+ people)

This is a simple, but fun game in which one person secretly "cyber stalks" a friend. If both people use Gowalla, it's even better. This game can be played on Foursquare, but with limitations because of the photo taking and especially because Foursquare doesn't require you to be in a location to check in there.

So you begin stalking your friend. The object here is to check in at the same location as your friend without being seen (physically) by them. There are a lot of ways to accomplish this. You could hide in the crowd, wear a disguise, or just pop in long enough to get the stalk and then get out before being seen.

The best way to play this game is to stalk your friend, wait for them to check into the location, then check in right after them with a photo of them attached to the check in. If your friend is paying attention to his or her Gowalla stream, they'll notice right away and begin looking for you. That's where the fun starts.

Last Check In Buys

(2+ people)

The name explains it all. This game is a simple "last one there's a rotten egg" dash for the prize. Upon leaving work or arriving at a bar, coffee shop, restaurant, etc., you can send a shout to your friends to tell them where to meet and mention that the last one to arrive buys (first round, meals, dessert, etc).

The location can be literally anywhere, like a BBQ at your house (last one to arrive has to make the beer run) or at the coffee shop. This game is easy to set up and can be fun.

Steeple Chase

(2+ people, several locations)

This is like a real-life steeple chase, but played through Gowalla or Foursquare. The object is simple: set up a route with check points and then see who can go through the route the fastest. With Gowalla Trips, this is especially easy and the spots to check in can be just about anywhere everyone is willing to travel to and from.

The best part is, the geo-location app is the judge as everyone can look to see who checked in where, when, and whether everyone made all the checkpoints. The game can be played with all participants going on the steeple chase en masse or one at a time with a timer.

Played from a bar, club, coffee shop, etc., this game is a great way to relay players in pairs during a party or get together.

Virtual Snipe Hunt

(3+ people, 1+ location)

This is a fun game to play as a prank on someone for their birthday, special occasion, or just to mess with them.

A snipe hunt is any game where the object is given to the players as something to find, but one or more players realize that the "snipe" doesn't really exist. With Gowalla, this can be played with items and with Foursquare it can be played with non-existing badges of honor.

For the game, choose a location (or the location itself could be the snipe) and tell everyone that the object is to find that location or to find a specific object at that location. Perhaps there's a virtual left boot at Anna's Pet Store and only the lucky person who finds the store and checks in first will find it. Or maybe Anna's store doesn't exist and players will spend all their time trying to locate something that isn't there at all.

The game works best with larger groups where only a couple of people are not clued into the true object of the game. Which is why it's best suited for birthdays and similar occasions.

Gowasquare Scavenger Hunt

(3+ people, multiple locations)

This is a variation on the steeple chase theme wherein each location has an objective as players visit several locations to gain objects. It has the added bonus of being a game that can be played over time rather than live.

A specific list of locations (like a Trip) as well as real or virtual items to be gathered while visiting all or most of those locations. From there, players visit each location in whatever order they'd like and attempt to retrieve the items on the list – or at least as many items they can from the list.

Hints as to what's to be found are better than just a straight list and make the game even more fun, but will require some kind of moderator or a way to list clues separately from the apps themselves. Since Items in Gowalla are constantly changing locations, this can make the game even more challenging as specific items can be "planted" and included on everyone's list. Rather than having people return the items, part of the quest can be to drop certain items at certain times/locations or by giving point values to items so that players will drop lower point items in order to gain higher point ones.

The ways to vary this game are nearly endless.

Spot the Stranger

This can be played with two people who may or may not know each other. Most of us have friends online we have never met in person. This can be a way to do that. Or disguises can be worn so that neither person will (hopefully) recognize the other without some detective work.

All kinds of role playing options are also available through this game. Fun scenarios or fantasies can be played out as can a bit of cyber stalking and surprises.

Some options would be for two people who know one another and use Foursquare or Gowalla to meet at a location, but each is required to bring someone the other doesn't know and arrive separately so the new people have to find their new friends.

If Sandra tells Bill to add her friend Tasha to his friends list and then sets up a meeting for them, but doesn't tell either of them what the other person looks like, this game will work as a match-making/blind date enabler as well.

Who Dunnit Adventure

(large groups)

This game is much more elaborate than the others suggested in this book, but it can be a lot of fun. Think of it as playing Clue without the board and instead with real live people. First, an objective should be named: finding a missing object, finding a suspected killer, or finding out where the ultimate goal is even located are all options here.

Once that is done, individual "characters" for people to play should be created and each should be given a location or locations to haunt as part of their job. Alternatively, everyone can be looking for one or more objectives while staying in character and finding clues through what other players might know.

So if there are three separate objectives, but each player only knows of one of the three, the game can be played with no referee and lots of players. Let's say that one player is looking for a murderer while another is looking for a thief and still another is looking for a stolen item. The thief stole the item and in the process had to kill someone to get away without being caught. Hence, all three are tied together, but probably don't know it.

As the players look for their objectives, each has clues that all lead to the same point, so as they meet and discuss, they can learn or trade clues and probably not even know that they're all actually looking for the same thing (e.g. the murderer who stole the diamonds).

Game Possibilities

The game possibilities with Gowalla and Foursquare are limited only by your imagination. Most traditional party games can probably be adapted to work with the apps – and many will be improved by doing so.

So have some fun! It's not just all about the check-ins!

Chapter 7

Security and Safety Tips for Foursquare and Gowalla Users

Privacy and security concerns are inherent in any public software you use online, especially social networking and even more so with geographic-location-based apps. Both Foursquare and Gowalla have the potential to be security problems and using either can, in many ways, remove much of the privacy from your daily life.

While this sounds scary, it doesn't mean you should abandon using these games. Covering up the reality with platitudes, however, would give you a false sense of security when using these apps and that would only lead to trouble, so it's better that we're up-front about this.

The first thing you should understand is that when you use these applications, you are exposing your current whereabouts whenever you check in. This is broadcast to your friends, colleagues, and potentially anyone who wants to know this information. Normally, while this might occasionally be something you don't want people to know, it is no different than posting your favorite hot spots and pictures of yourself to Facebook or Twitter.

Understanding how you can ensure your own safety and security is what this chapter of the book is all about. Of course, the technology changes quickly and no one can provide for your perfect safety and security, but you can minimize most problems by following the tips in this chapter.

The First Rule of Safety

For Gowalla and Foursquare (or any geo-location network), the first rule of safety is to understand that these applications don't just tell your friends and network what you're doing. They physically locate you in the real world.
This is very different from a service like Twitter where announcing that you're "going to Java Jim's for coffee" may or may not tell those on your network where you are. With Foursquare and Gowalla, this literally tells them (via GPS coordinates) where you actually are.

This should make it obvious why real common sense towards safety is very important with these networks.

Never Assume Security

The second cardinal rule to never forget is that security is not "someone else's job." Foursquare and Gowalla are very well-made software platforms and both companies have gone to great lengths to ensure the security of your data and network.

But no safe is un-crackable and no software is completely secure. That is true whether you are looking at your bank's online website or MySpace.
Think about this: the same day that Apple's new iPad tablet was released to the public, someone figured out how to hack into it and gain access to the very core of the machine. This meant that they could run anything they wanted (whether the iPad was theirs or someone else's) and displayed how easily someone who really wants to be able to access your data can do so.

So keep your private information to yourself, as you should with all Internet communications. Just as you wouldn't email someone your Social Security Number, you really shouldn't send your daughter's school schedule via Fousquare either.

Your Home is Your Castle, Not a Spot or Check-in Point

One of the most obvious security realizations is that your home is probably not the best place to list as a check in location for either Foursquare or Gowalla. First, this tells everyone on your network (or who gains access to it) where you live. By inference, that tells them, whenever you check in somewhere else, where you aren't.

So when you check in at Molly's Cafe three miles away from your house, someone with ill intent now knows that you aren't home and that now is a good time to pull up with that "delivery van."

Likewise, your friends and family should be aware of this and their homes should be considered sacrosanct as well. If they have made their home a spot on one of these games, then they've consented to allow you to check in there. If they haven't, it is definitely not very considerate for you to "discover" it and make it a spot for them.

This is probably the most hard-and-fast safety rule you can follow with both Foursquare and Gowalla, or any other location-based service.

Know Who Your Friends Are

Another extremely important rule that applies to geo-location application networks is to know who it is you're making friends with. Unlike Twitter or Facebook, networks like Gowalla and Foursquare are much more personal because of the physical likelihood of meeting.

If someone you don't kow or can't identify asks to be your friend on a geo-location-based network, you are only putting yourself in danger by accepting their friendship request. Find out who they are before you accept their invitation or reject it out of hand. If they really know you, you will get a phone call or an email explaining who they are.

Just Use Common Sense

The last word in security on these geo-location applications is to just use common sense. If it doesn't seem like a good idea to broadcast something or check in somewhere, then don't. If "founding" a new Spot on Gowalla is making you a little nervous, then don't do it.

Do you really want to be the permanently-listed Founder of Jiggly's Strip Joint knowing that your boss is a devout Christian woman? Never mind what your wife might think?

Use common sense with your security as well. If broadcasting that you're on vacation in Germany and your house is completely empty and defenseless seems like a bad idea, that's because it is.

Finally, if you do meet new friends through Gowalla and Foursquare (which can be a big part of the fun), make sure you're always doing so in very public, well-populated places. If you arrive at a location to meet someone and it doesn't look right, leave immediately. You can always come up with an excuse later as to why you were a no-show.

As always, Safety First and have fun!

Chapter 8

Capitalizing on Foursquare and Gowalla to Promote Your Business

This will be one of the largest chapters in this book – almost a whole book in itself, really. Why? Because Foursquare and Gowalla aren't just games, they're the latest social networking applications on the Web that are not only gaining popularity very quickly, but that have huge implications for the small business owner and entrepreneur.

Unlike Twitter, Facebook, MySpace, and other social networking apps before, geo-location applications are linked to the physical world. This opens up a huge barn door for innovative and creative business owners and marketers.

If you've read this far into the book already, you've no doubt had a spark light in the back of your mind regarding this. Well, let that spark turn to a flame because the newest brush fire of the business frontier is about to roar through.

It's likely that Foursquare and Gowalla could be the biggest marketing opportunity for small and local businesses to ever hit the Internet. Let's see how you can get in on this now, while it's still new and fresh.

Are Foursquare and Gowalla Large Enough to Market Towards?

Marketing on small, niche, low-traffic websites can pay off if you happen to cater to the particular niche market that is presented. Foursquare and Gowalla, however, are not really niche markets in that sense and so any marketing person has to ask themselves whether it's worth utilizing them (yet).

This is a valid question and it can be answered simply: maybe.

Foursquare has passed the one million users benchmark, world-wide, and so if your businesses is in an area populated by smart phone users, it's likely that you will find benefit by using Foursquare. Gowalla, similarly, has hundreds of thousands of users (and growing fast), so the same rule applies. Getting in early has other advantages, since most businesses in your area aren't likely doing much (if anything) with these geo-location apps yet. Being one of the first gives you more than just credibility, it can also become branded in people's minds: you can become the Foursquare hot spot of your area.

The largest benefit to your business will be the online word-of-mouth that these apps will give you. Those who currently use them tend to be engagers and centers of sizable hubs of user networks, which means every time they mention you, you get advertising. There is no more powerful marketing than personal recommendations and word-of-mouth.

Learn From Successful Examples

Gowalla and Foursquare haven't been around very long, but already major product and business brands are taking advantage of them. A handful are doing so with great success. Let's look at three very different businesses and how they're using the unique opportunities that these geo-location apps offer and fitting them into their unique business model.

Of course, your business might be different from these three, but these examples should give you a good idea of what is possible. The idea here isn't to give you a template to copy, but instead to get the wheels turning in your creative mind so you can see what might be possible for your business.

After these three examples, we'll look at other ideas that could be used to promote a small, local, physical, or just about any other business.

The Resort Hotel Model

A resort hotel (or any similar enterprise) is a little city in itself and thus can become a whole host of geo-locations for check in opportunities. Wynn Las Vegas realized this quickly and their electronic marketing personnel immediately set about creating interaction with customers in and around their resort.

Locations within the hotel and its surrounds were made into Spots and each was populated not only with its own check in location, but with helpful hints about it to show connected customers what might be possible. Check-ins are encouraged and some locations have special deals or offers available only to Foursquare users.

Wynn

The goal at Wynn, however, was not to be first, but instead to enhance the vacationer's experience via Foursquare. It has worked very well with a lot of positive feedback from customers who were surprised to see a huge, multi-million dollar company like Wynn already using something as new as Foursquare.

The Small, Local Business

Whether retail- or service-oriented, the local business can definitely use Foursquare and Gowalla to gain new customers, enhance existing customer's experience, and reward repeat customers. A great example of this is a company called Tasti D-Lite in Nashville.

This alternative dessert shop has a full-blown social media loyalty program in which Twitter, Facebook, and Foursquare users are rewarded for their loyalty and interaction through the social media networks. They call their program TastiRewards. It all ties in with their online store, giving rewards for repeat shoppers and those who talk about Tasti D-Lite to their friends.

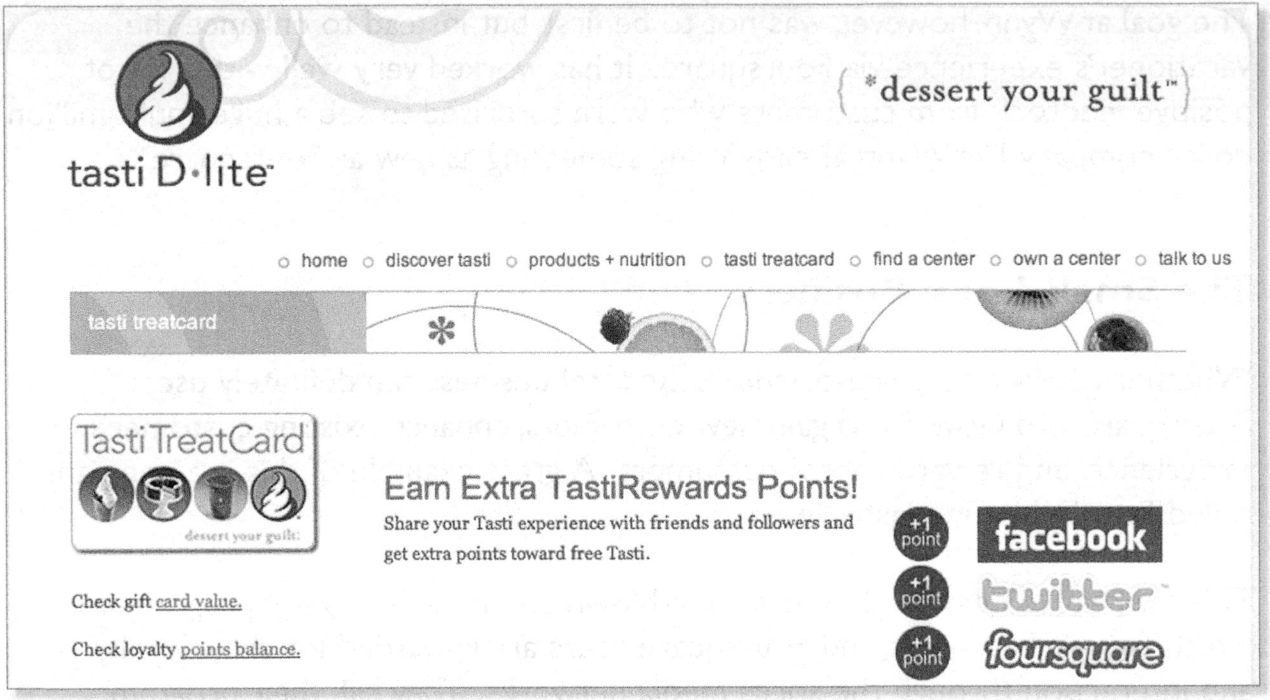
tasti D·lite
*dessert your guilt
home
discover tasti
products + nutrition
tasti treatcard
find a center
own a center
talk to us
tasti treatcard
Tasti TreatCard
dessert your guilt
Earn Extra TastiRewards Points!
Share your Tasti experience with friends and followers and get extra points toward free Tasti.
+1 point
facebook
+1 point
twitter
+1 point
foursquare
Check gift card value.
Check loyalty points balance.

The company has had so much success with it, they've spread their program to 30 other locations around the country.

Rewards are based on check-ins, social network shout-outs, and similar activities on the part of the customers. The more complicated Tasti D-Lite idea can be scaled down for smaller businesses, or scaled up as Tasti has done for a national franchise.

The Magazine As Shopper's Guide

In this example, a local magazine that includes a local review guide has captured the power of geo-location to promote themselves by offering free information to users in their city. Drawing on its archives of reviews of various hot-spots around the city, *Lucky Magazine* has added Tips in Foursquare to many area hot spots and is working to do this nationally with over 600 locations already reviewed in Foursquare by the magazine.

foursquare
Find places, people, tags
SEARCH
ME
HISTORY
STATS
FRIENDS
Currently in Boca Raton, FL
SHOP YOUR CITY
with Lucky magazine and foursquare
Lucky's Guide
1. @ Virtu: Beautifully crafted home goods and accessories make this spot a favorite general gift store of sorts. (2 weeks ago) [Link]
I've done this!
Add as To Do
Ignore
2. @ Erin Gallagher: Colorful gems are the focus of this designer who creates her own collection of pieces but encourages you to DIY at her in-boutique gem bar. (2 weeks ago) [Link]
I've done this!
Add as To Do
Ignore
3. @ Jayson Home & Garden: We're not sure what we love most at this haven for all things home: the flea market, the garden, or the indoor furniture and accessories. (2 weeks ago) [Link]
I've done this!
Add as To Do
Ignore
FOLLOW NOW
Lucky Magazine on foursquare - Your Mobile Shopping Guide
Shop where Lucky editors shop and get exclusive insider tips and deals!
Follow us and you'll be alerted to the best spots for jeans, vintage clothes and accessories, and hard-to-find indie labels. Visit our favorites and you'll unlock the Lucky Elite Shopper badge.
Luckymag.com

The idea began simply, though, and is scalable to any level. First, the *Lucky Magazine* website started a Foursquare Guide called the Lucky Mobile Shopping Guide and expanded from there. They incorporate reviews not only in Foursquare Tips itself, but also in the magazine's physical publication, by pointing out Lucky Mobile Tips in new reviews and listings.

The magazine is not only offering helpful information to geo-location users, but also branding itself to potential new readers who enter locations where Lucky has left its mark.

The Key to Successful Marketing With Geo-Location

The true key to marketing with geo-location apps like Foursquare and Gowalla is value. This can be value like you would provide in a blog, a website, or some other online venue. More likely, though (in order to really take off), it will be physical value.

Since these apps are based in the real-world, your value will be more successful if you give real-world payoffs for using the apps at your location. So if you own a diner and have set it up as a Foursquare and Gowalla hot spot, consider putting up a small sign that offers free coffee or free pie to those who check in when they visit.

Of course, "the mayor eats for free" (or gets free haircuts, or gets a free car wash) is always popular, but creativity can enhance this idea even further.

Coupons and similar incentives are the fastest-growing and most effective way to add value to your geo-location marketing. The burrito chain Boloco in New England has a Mayor of Boloco Card for each of its 16 locations. When someone becomes mayor for checking in, they receive the card and get specific perks. Plus, when the office of Mayor changes hands, Boloco makes a big deal of it and has a "handing off ceremony" where the old mayor gives over the card to the new one.

Another Case in Point: The Nets and Gowalla

Vaynermedia, a social media brand marketing company founded by brothers Gary Vaynerchuk (known for Wine Library TV) and AJ Vaynerchuck (the tech behind Wine Library TV). The New Jersey Nets are clients of Vaynermedia and set up a test run for using Gowalla to increase crowds at games.

Vaynermedia then gave away 250 pairs of tickets through specially-made Drops in Gowalla, which they positioned randomly over several locations throughout the tri-state area including parks, sports bars, gyms, etc. The virtual tickets could be redeemed at the box office to get into the actual Nets game they represented.

All of the tickets were collected and more than 15% of them were redeemed for actual seats in the game, which is a huge payoff for a virtual giveaway for real items. The Vaynerchuks see this as a huge opportunity that can be taken advantage of by virtually any business to replicate that success.

One of the authors of this book (Herb) found Nets tickets himself!

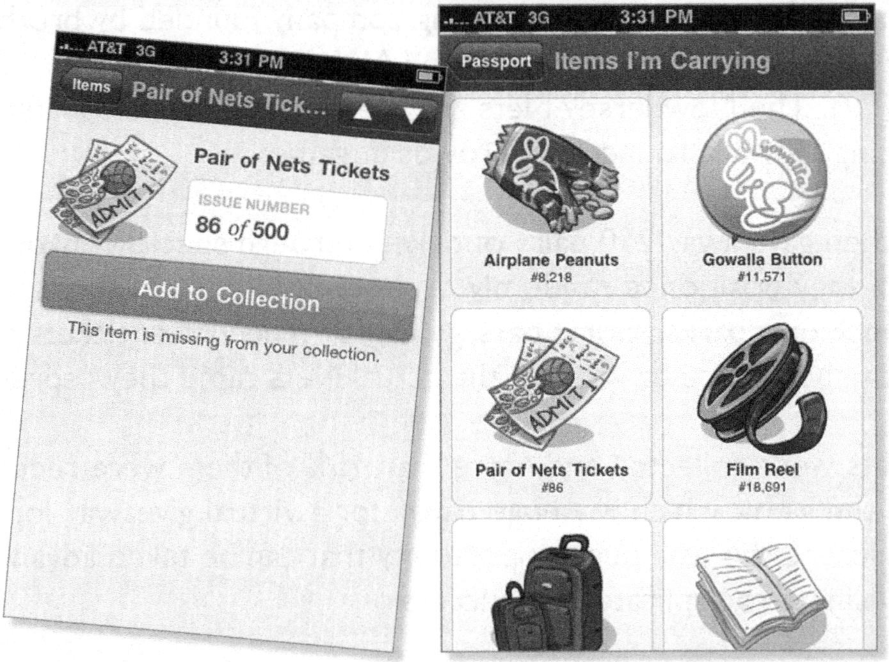

Creativity is the Core Attribute

The best thing you can do to use geo-location in your marketing is to be creative with it. A chain of thrift stores could have scavenger hunts hosted on their Gowalla accounts or a group of restaurants along a block or two could host a "walking wine tasting" on Foursquare.

Adding value by promoting creative ideas to entertain, educate, or otherwise engage your customers (and prospective customers) is the most effective way to market just about anything, including your business, on geo-location apps.

For many businesses, games are a fun way to get creative with geo-location. See our games suggestions in Chapter 5 for ideas you can adapt to promoting your business.

Promote Your Business Effectively on Both Foursquare and Gowalla

Using one or both apps can benefit your business if you're willing to take the time to engage and to make the check-ins of your customers and prospects unique. The way they're greeted when they check in, the bonuses they might receive if they check in often, or the coupons they receive when checking in to tell their friends are all ways that this can benefit the business.

The key is that you need to engage your customers and clients when they participate on one of these networks. Keep your Foursquare or Gowalla account for your business active, use it to interact with customers, and more.

Say a person walks into a pub and sends out a shout on Foursquare saying they just stepped into your pub to order a quick pint of Guinness and then walks to the bar and the bartender says "you the one who wants the Guinness?" That builds an instant, positive connection with that customer. They're going to talk about it.

The Difference Between Marketing on Foursquare and Gowalla

The biggest difference you'll find when marketing towards Gowalla versus Foursquare is in the way the two games work. Foursquare emphasizes the number of check-ins, with most rewards coming to those who check in the most often. Gowalla, on the other hand, usually randomizes rewards based on location and rewards available.

This makes Foursquare the more customer-loyalty-based and Gowalla the more flexible of the two for marketing a business.

With Foursquare, you will most likely be rewarding customers for repeatedly coming back to check in at your location. With Gowalla, you'll likely be giving away free items, coupons, and other incentives on a more or less random basis.

In either case, both services are a great way to market your business.

Promoting a Virtual Business with Geo-Location

Even businesses that don't have a physical address or brick-and-mortar presence can use geo-location apps to promote themselves. Of course, many times these businesses can see great results from other, more virtual social networks. That doesn't mean that geo-location is closed to them.

A virtual entrepreneur still has physical contacts in the real world. Perhaps you can establish value by promoting your business through a Gowalla account which you then use to leave feedback and hints about places you visit as part of your business. A Web designer might spend a lot of time at an art store, an office supply store, etc. and have recommendations for people who might also visit those places.

Again, the idea is to be creative with your marketing. The opportunities on Gowalla and Foursquare are literally wide open and mostly unexplored. Watch for changes in the networks themselves to come and give more advertising and marketing options as well.

Geo-Location as a Workplace Tool

Yes, games can be played at work and actually enhance productivity. We mentioned earlier that college campuses now have tours and hot spots popping up that students and faculty are adding to Fourquare and Gowalla to enhance the experience of new students, including Harvard and UNC Charlotte.

Some universities have built campus-approved Trips in Gowalla and some businesses are now using Foursquare and Gowalla to create tours of their facilities. There is an unofficial Gowalla Tour of the Google Campus in Silicon Valley, the City of Boston has a municipal tour of its governmental buildings.

The Zimmerman Agency in Florida uses Foursquare for both fun and productivity with their campus being largely discovered and about 35% of employees using the app. Employees use it to locate one another, but also to locate key customers and suppliers in the city.

The geo-loc apps can also be used for employee recognition and morale building. Playing simple games, giving special awards or perks to the Mayor or those with specific badgets, and otherwise using the apps as part of an overall team-building effort can pay off.

Chapter 9

Looking at Other Geo-Location Services

Of course, the geo-location marketplace isn't populated by only Gowalla and Foursquare. There are literally dozens of others. There are a few that are more prominent (in terms of size or potential size) than even the two we're emphasizing in this book.

Those prominent services merit a mention so you can be familiar with them as well. Some are more conducive to business marketing while others are more socially-oriented or more game-play centric.

In this chapter, we'll look at some of these other geo-loc services and give you an idea of how they work and why they may or may not be good for business or fun.

Facebook Check-ins

Yes, the big boy of social networking is now getting into geo-location. Facebook introduced a location-based "check in" feature this year. While limited in use, for now, the addition has big ramifications for the social networking – geo-location scene.

The service could potentially tie 300 million or more people into geographic-centric uses. It also means that some Facebook add-ons and games could also become geo-centric.

For now, Facebook is mainly a social networking site and most businesses will see most of their marketing activity happening on that end. Watch for some apps like Foursquare or Gowalla to enable automatic check-ins through Facebook as well as their own networks, though. This could potentially open your geo-location marketing to a new dimension.

Goby

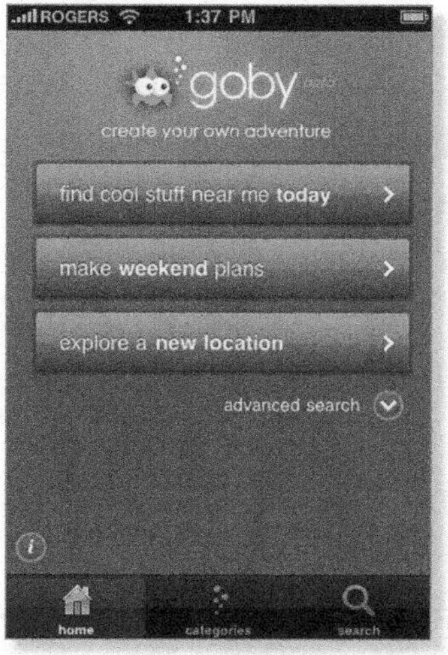

This is a sort of yellow pages for geo-enabled phones. It's a great service that allows you to find just about anything based on where you're located at the moment.

Emphasis in Goby is on activities like hiking trails, dog parks, bicycle rental stands, and so forth. It's a great vacationer's app, a fun "weekend with nothing to do" resource, and more.

For the business owner, if your business fits into the kind of thing that Goby caters to, then you will find it useful for marketing, but in a limited way. Once you're entered into their database, your only concern will be updating any information as needed.

Google

Although Google (or its new Twitter-like interface, Buzz) hasn't gotten into geo-location per se, the search giant is a major player in geographic Internet data. Google Maps are used in most apps that access mapping and Google compiles a huge amount of geo-centric data.

Look for services to pop up that surround or utilize Google's database beyond what is being done now. Simple maps will probably soon become more three-dimensional (and are in some areas).

For its part, Google has also been making inroads towards gathering data from Web 2.0 and the Social Web, pulling this together for use in various ways. Watch for future innovations in how that data is crunched and delivered. Especially to geo-location apps as they mature.

For the business owner, this can mean a wealth of marketing information and access right at your fingertips. It will also mean more and more accessibility for your potential customers as GPS become more common, which will mean making yourself seen will be paramount in the local Web.

MyTown

Located at Booyah.com (its parent company), this is actually the longest-lived geo-location app with the largest userbase of any, including Gowalla and Foursquare. It is basically a real estate game with users buying and selling actual plots of land virtually.

When other people visit your shop and check in, you collect "rent" in the game. The more frequented a location, the more rent you collect and the more valuable the land is if you decide to sell it. Think of it as monopoly on the smart phone and in real life.

For the business, there are a few uses for this app in marketing. The most obvious is to give incentives to the property owner, for those who own certain items in the game (these are similar to badges or drops), and for those who promote check ins to your location.

For example, a Quiznos location in Seattle offers a $2 coupon to anyone who checks into their location at the cash register using MyTown.

Whrrl

Combining social networking with real-life friendships and geo-location, Whrrl is one of the newest of the geo-loc apps out there. In Whrrl, users can join "Societies" based on their interests, where they've been, and who they know.

If the local coffee hangout has a "Regulars Society," then a person who checks in there with Whrrl can opt to become a member. This opens up new avenues for other recommended places to join as other members of the Regulars can list their favorites. If your interests match theirs, you'll likely find new places to visit. Less a game than it is an innovative social networking application, Whrrl runs on iPhone and Blackberry devices.

TheDealMap

This geo-location app is a discount-finder that uses your current location as its basis for finding you deals. If you're in Topeka, Kansas, for instance, and want to find a good deal on a new t-shirt to show the world that you've visited Topeka, then you can use TheDealMap to find the best deals on local souvenirs.

TheDealMap scours most of its listings from the Web, but your business can also list itself there with any specials or promotions (even if they're TDM-specific) you want to include. It's not really a socially-enabled app, so there is no reason to spend a lot of time with it to promote your business.

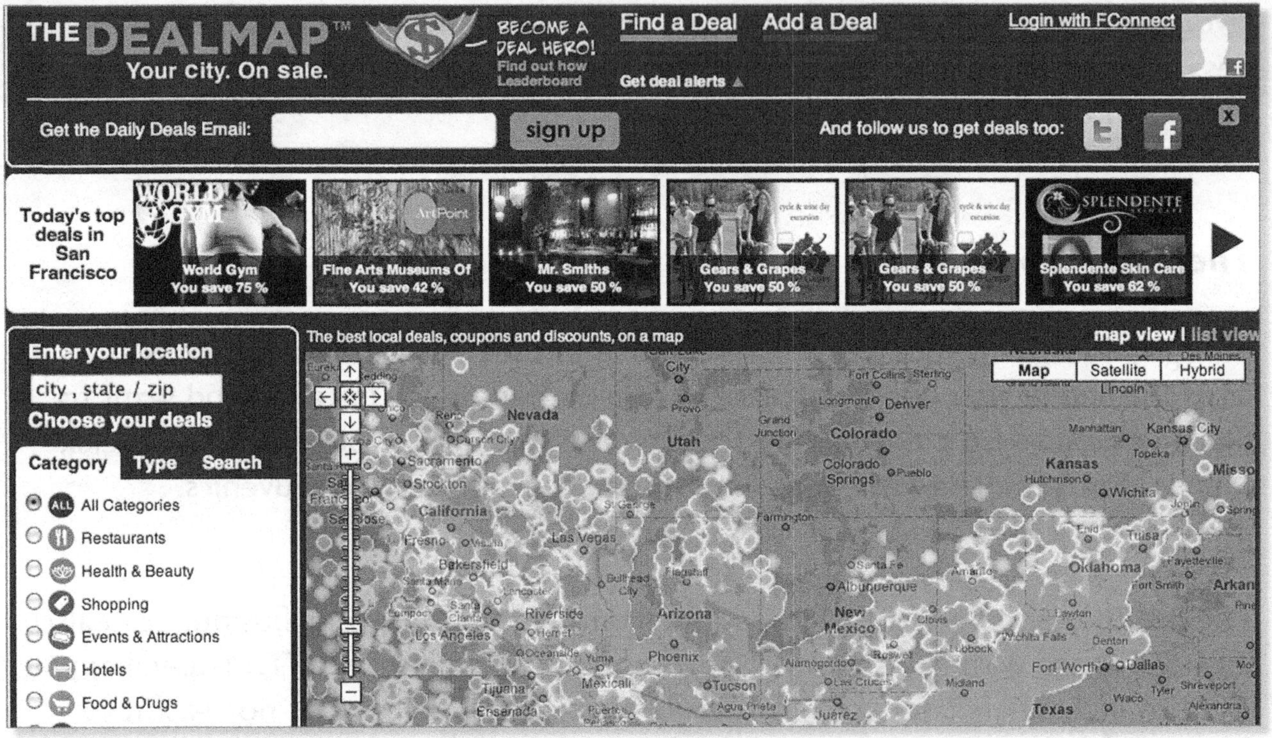
THE DEALMAP™
Your city. On sale.
BECOME A DEAL HERO!
Find out how
Leaderboard
Find a Deal
Add a Deal
Get deal alerts
Login with FConnect
Get the Daily Deals Email:
sign up
And follow us to get deals too:
Today's top deals in San Francisco
World Gym
You save 75 %
Fine Arts Museums Of
You save 42 %
Mr. Smiths
You save 50 %
Gears & Grapes
You save 50 %
Gears & Grapes
You save 50 %
Splendente Skin Care
You save 62 %
Enter your location
city , state / zip
Choose your deals
Category
Type
Search
All Categories
Restaurants
Health & Beauty
Shopping
Events & Attractions
Hotels
Food & Drugs
The best local deals, coupons and discounts, on a map
map view | list view
Map
Satellite
Hybrid

There are a lot more finds on TDM than you'll find from just about any other source. It's a useful application for the geo-phone user.

Twitter Check-ins

Twitter is also getting into the geo-location game with a geo-based check in addition. You can now tag your tweets with your location and even make that tagging automatic if you're using Twitter from a geo-enabled phone.

It's not likely that much more will come of this than what Twitter's always done, but you should never say never. Twitter is opening the geographic portion of their API to developers, so just about anything could happen now.

Like Facebook, watch for Gowalla, Foursquare and others to allow dual-checkins for both their own app as well as Twitter. This, again, opens a large new market for social network marketing for the business owner.

Yelp

One of the largest location-based apps out there, Yelp is a consumer-centric ratings site for businesses. In basic form, the app works by allowing you access to the Yelp site and database. You can browse businesses by location and category (business type) and if you've used the business, you can leave feedback on how they were for other users to read.

Yelp has had its ups and downs, but is still a trusted source for a lot of Web-savvy consumers. Used on the smart phone, it's even more useful for on-the-spot hunts for the right place to do business.

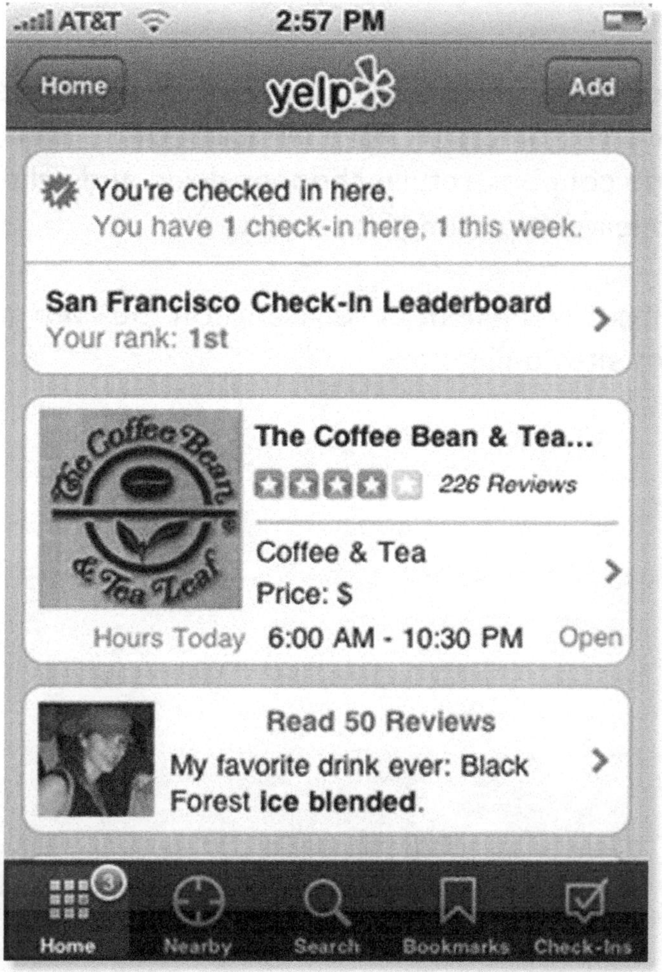
AT&T
2:57 PM
Home
yelp
Add
You're checked in here.
You have 1 check-in here, 1 this week.
San Francisco Check-In Leaderboard
Your rank: 1st
The Coffee Bean & Tea Leaf
The Coffee Bean & Tea...
226 Reviews
Coffee & Tea
Price: $
Hours Today 6:00 AM - 10:30 PM Open
Read 50 Reviews
My favorite drink ever: Black Forest ice blended.
Home
Nearby
Search
Bookmarks
Check-Ins

The business owner will want to make sure their business has a complete profile on Yelp. Encouraging customers to leave feedback there is another good idea. Many businesses use coupons, return-shopper deals, and other incentives to get their customers to leave Yelp ratings for them.

Yelp is one of the most prominent destinations on the Web and is one of the most trafficked consumer sites online.

Chapter 10

Location Based Apps

Whether your looking for a way to simultaneously check in to multiple geo location apps or simply need to find a place to pee, well there's an app for that! Despite being relatively new already a crop of new geo location apps have arisen to work with Foresquare and Gowalla.

While the following list only presents a few of the more popular apps expect the list to expand as the geo location craze continues. On a cautionary note some of the apps listed might not work depending on your mobile device. Additionally, on occasion API changes from the geo location companies can blackout an app for a period of time, and while the blackout might not be permanent it can be frustrating.

So go try these apps and enjoy.

Foursquare

Footfeed

Footfeed is the first native iPhone app that makes it easy to check-in to Brightkite, Foursquare and Gowalla simultaneously. Built from the ground up as a check-in aggregator network, Footfeed aims to bring together the check-in services so that you can check-in to many networks without the need to open multiple apps. No more wasting time checking in using multiple apps. No more check-in fatigue. Using Footfeed to check in is easy.

Simply create an account and link up your Brightkite, Foursquare, Gowalla and Facebook accounts. Then when you want to check-in, select your location and press check-in. The app will automatically match your Footfeed check-in to the location across the different services you linked up to your account and check you in simultaneously. And you will continue to earn your points, badges, mayorships, rewards and rankings from the check-in networks.

AT&T 3G
11:29 PM
6%
footfeed
Done
Gowalla
Username: Dennis Mink
Cafe Du Soleil
This is your 1st check-in here!
Foursquare
Username: Dennis Mink
Mayor status:
Poppa B. is The Mayor of cafe du soleil.
Checkin status:
OK! We've got you @ cafe du soleil.
This is your 1st checkin here!
+8 points
Travel bonus: 8 stops
Checkin
Places
Friends
My Profile

iPod
6:04 PM
THRILLIST
EAT
DRINK
SHOP
NEW YORK
Rivington St
Forsyth St
Orchard St
Ludlow St
Delancey St
Essex St
Casa Mezcal
This three-floor ode to Oaxacan culture features...
Delanc
Broome St
Grand St
Grand St
Google
NEAR
NEW
TOP
MY
FUN

Thrillist for iPhone

Check in from spots that don't suck directly from the spankin-new Thrillist for iPhone app, complete with seamless Foursquare integration. A supplement to our free daily email, Thrillist for iPhone narrows down the dizzying scope of your burg to deliver you only the handpicked nearest, newest, and most popular – from five-star steaks to two-dollar tacos, rooftop clubs to underground speakeasies, and bespoke shirts to kicks on such deep sale it's almost a five-toe discount. Stay on the cutting edge of your Foursquare friends by checking in at these dizzyingly sweet places they never knew existed.

To check in, simply select your currently awesome location, tap Share, and select the Foursquare entry of your choice.

Thrillist for iPhone covers: Atlanta, Austin, Boston, Chicago, Dallas, Las Vegas, London, Los Angeles, Miami, New York, Philadelphia, San Francisco, Seattle, & Washington DC.

Spinlet

Spinlet is a music network with Foursquare music Check-Ins. Check-In to a venue with your favourite tunes. See who, what and where on Spinlet real-time map view. Spinlet is available for iPhone and Android devices. Release June 2010.

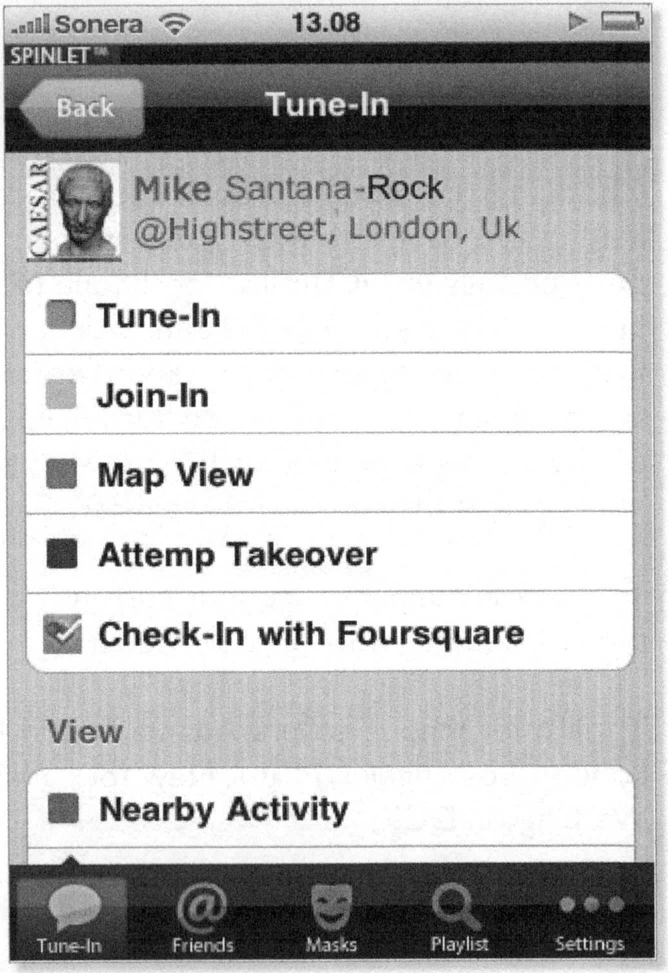

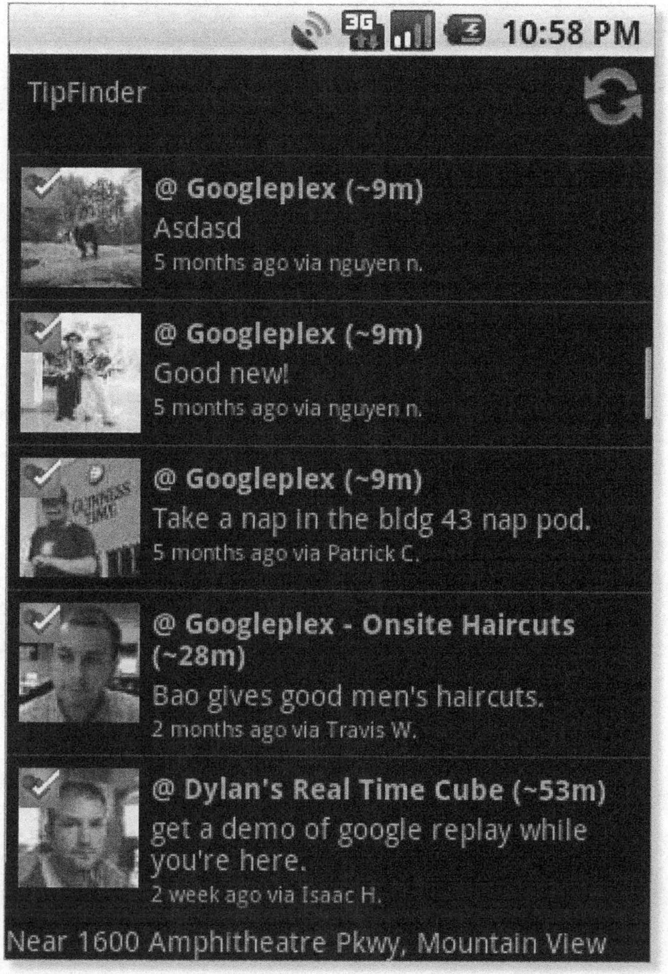

TipFinder

An Android application to help you find the best places to eat and drink by utilizing Foursquare's Nearby Tips. No account required, just start using it! You can download the app by searching "TipFinder" in the Android market place.

Fiddme

It's a restaurants guide... it's a friend-finder... it's a game... it's Fiddme! Fiddme aims to be a local visual guide to the culinary wonders around you exposing new to new places and friends in your area. It's super easy to checkin while you post a new bite.

iSlide Camera

Supercharge your check-in with an instant, beautiful photo gallery. ISlide Camera lets you snap photos and share them instantly with anyone – a Blackberry, an iPad, and more. Amaze your friends as your pictures roll in, in real time, all powered by a single tap.

Effortlessly create an entire slideshow, customize its design, and check-in to Foursquare in one tap.

iSlide lets you create slideshow, on the go, while checking in to a location. Use the camera while shopping, attending events, partying, traveling, or just hanging out.

Carrier
12:51 PM
The Green Carnation
Portland Pl
A4200
A400
B506
A40
A5204
A4201
B406
A401
B401
B404
New Bond St
Regent St
Bruton St
Drury Ln
Clear
Done
Featured
Shops
Dining
Cafés
Nightlife
Fun
Sights
Sports
Travel
Outdoors

Mappr

Foursquare & The World Cup = Mappr!

Our iPhone app lets you see Foursquare places on the map, prioritized in a way that reduces clutter and surfaces the best in any map view. Zoom out to see the top places in the whole city, zoom in to see everything right around you. Filter on category to find exactly what you are interested in.

We've taken the idea of checkins and added a bit to it. With Mappr, when you check in, you can also do so as part of a group.

With our World Cup edition, we have created groups for each of the teams. So whenever you are enjoying a game, or visiting a place that well represents your team, check in with that team to let the world know about it

TwentyThree

Tired of looking up your friends on two or three different networks every day? Tired of checking in on all those networks every time you go out? TwentyThree is made for you!

TwentyThree is an iPhone and iPad app that aggregates the position of your friends on all your social location networks and let you see them where they last checked in. But that is not all. TwentyThree also sports a native check-in module which will let you publish your location with all your accounts easily.

The list of supported services currently include: Foursquare, Gowalla, Brightkite and Twitter.

Edit
Friends
I'm starved!
Mark
Anaheim Hlls, Anaheim
6 hours ago
Keep dropping baby, $2.99 per gallon. (@...
Anthony
McDonald's
7 hours ago
lunch with Max.
Iain
Rotherfield, East Sussex, UK
8 hours ago
Ok that's enough experimenting with the BB...
MG
Meatpacking District, New York
8 hours ago
Working at the Standard for a bit then movin...
Chris
Starbucks Coffee
21 hours ago
Robert
Willow Garage

Features:

- **Map view:** see where all your friends are in one glance in the embedded map view.
- **Unlimited accounts:** TwentyThree accepts an unlimited number of accounts of each of the supported services.

- **OAuth:** TwentyThree will not store your credentials but use OAuth on all services that proposes a functional OAuth interface (currently Twitter and foursquare).

- **Friends merge:** TwentyThree will automatically merge contacts it recognizes as identical and propose an intuitive interface to merge the ones it could not identify.

- **Place merge:** The check-in module will also try to automatically merge identical places found on your different networks.

- **Twitter support:** Even if you do not often use a geolocation service yourself, you can follow the check-ins published by your friends in your usual timeline.

- **iPad support:** TwentyThree is an universal app which you can run natively on all your devices.

- **Built-in help.**

Pee*Free

Pee*Free helps with urgent matters while traveling. Display public restrooms (WC) on a map. View accessibility, fees and cleanliness. The database is community driven with locations from http://foursquare.com.

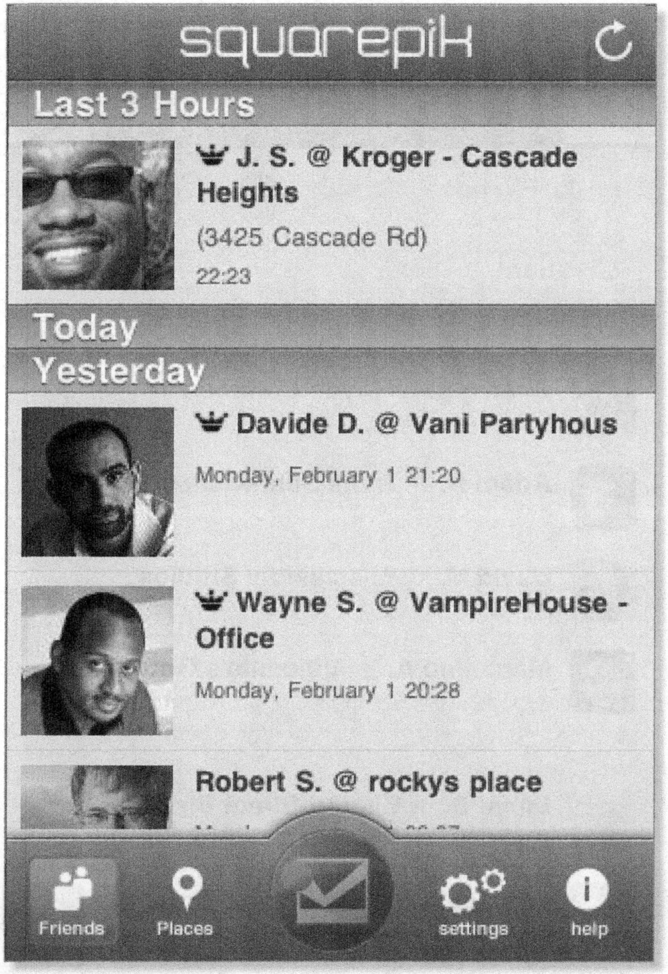

SquarePik

SquarePik is an easy way to add photos and videos to your Foursquare check-ins. It will also show you all the photos and videos that were dropped at a venue.

Kickball

A new way to play foursquare. Same login, friends, places and badges in a new beautifully-designed interface with some great new features. We simplified everything down to what matters most—friends and places.

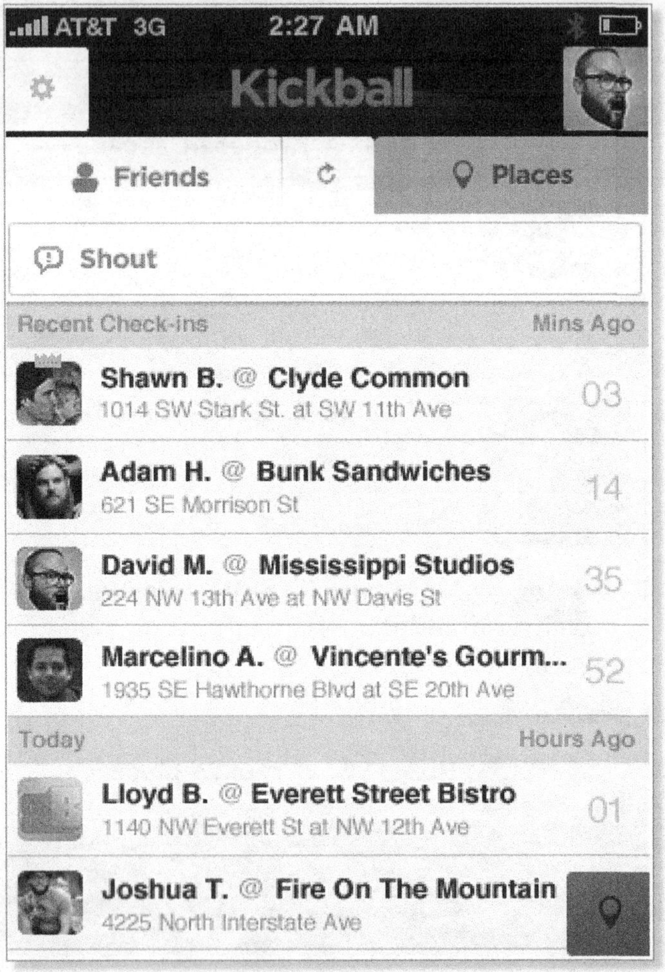

Gowalla

The Gowalla API enables tons of fun stuff. Here are a few free ideas:

Indiana Jones Map — Convert your road trip into an animated, epic adventure.

Sidebar Badge — Show off your top ten spots with a sidebar badge for your blog.

Conference Games — Encourage conference attendees to explore the city and get to know each other by tracking their check-ins and rewarding those who get off the beaten path.

LunchVote — Can't agree with your co-workers on a lunch spot? Stuck in a rut with the same five restaurants? Use LunchVote to decide on a nearby restaurant and track your history.

Gowallahashing — Inspired by XKCD's ***Geohashing***, algorithmically pick a random spot every day for a given area, to encourage serendipitous challenges and meetups.

Gowalla Sniffer — Takes the guess work our of locating those hard to find items for your vault, often right under your nose. Sniffer lets you view Gowalla items both near and far; without checking into locations. Use your iPhone's location services, or enter a city/state/zip and away you go.

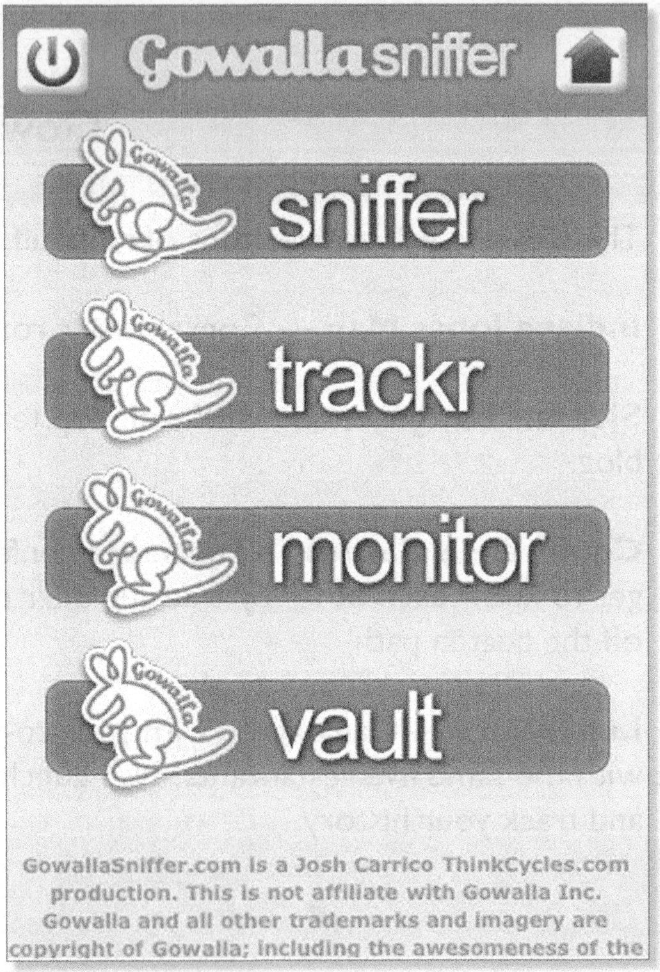

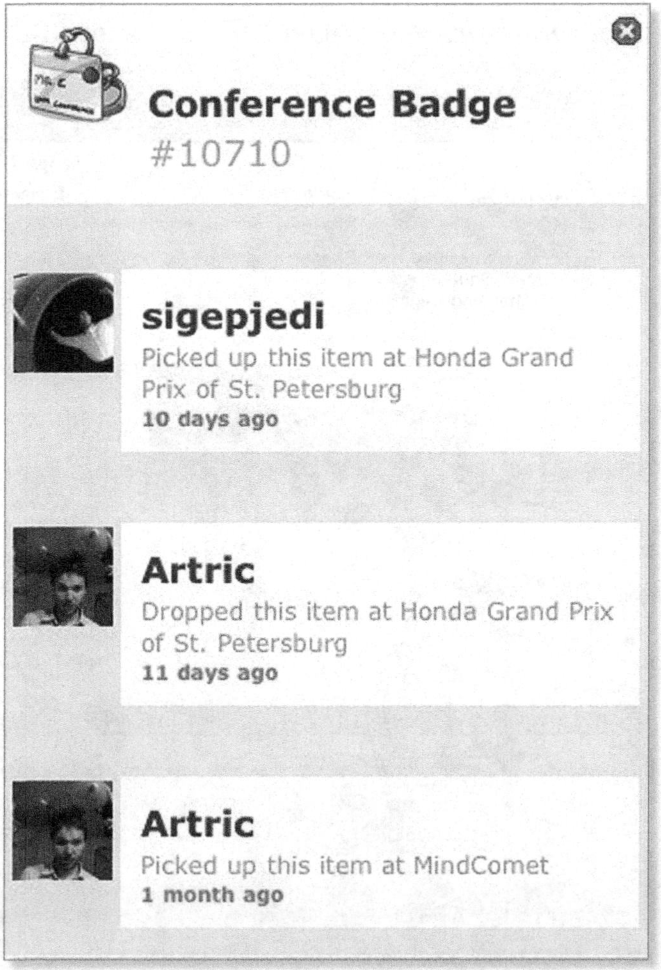

Let the sniffer monitor your locations and even your Gowalla friends! Keep tabs on Gowalla locations lets you know when items become available to grab for your vault or pack. Explore your friends packs to arrange a Gowalla item swap.

Gowalla Swag — Are you a big fan of Gowalla? Do you like the collect those Gowalla 'swag' *items?* Then GowallaSwag is the app for you.

With GowallaSwag, your life as a Gowalla scavenger hunter becomes real easy. It shows you all the nearest Gowalla spots with items. When you click on a spot in the spot list, all the items of the spots are displayed, something you can't do with the original Gowalla app, because you'll have to check in to see a spot's items.

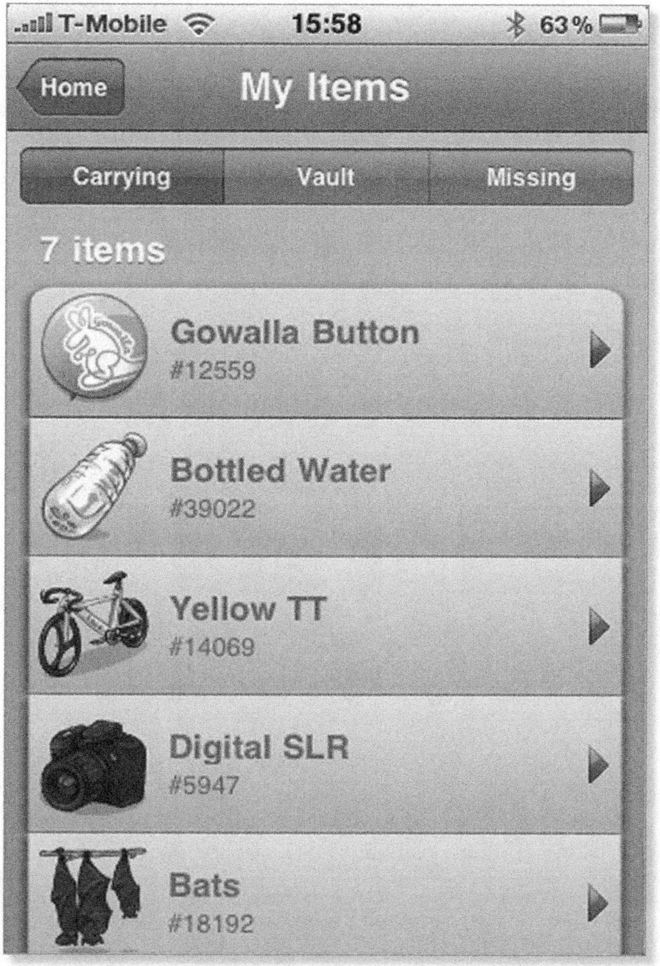

You can even check in at Gowalla spots within the GowallaSwag app, because the Gowalla Web client is built right in to GowallaSwag. For now, the Gowalla API is read-only, that is why we need the Gowalla Web client to check in at spots.

Chapter 11

Conclusions About the Location-Based Future

While location-based applications are hot right now, the future is not easily told. It is apparent that, in all likelihood, geo-location-based games and applications are going to be around for a while and will continue to gain in popularity.

The question becomes one of money, however. This is how most business and online applications have their fate decided and location-based apps like Foursquare and Gowalla are no exception. Currently, these apps have a relatively limited appeal because, in the end, there is little true incentive for users to continue to play.

Recent moves by both players show that the companies are aware of this and that soon incentives outside of the social networking games will be appearing.

These incentives will provide monetization for the applications as well, making them viable.

So while the future is uncertain, this is definitely the time to get in on geo-location. They will only become more popular in the next year or two and are ripe for exploitation for both fun and profit.

Made in the USA
Monee, IL
05 November 2026

67947704R10096